The Microwave Debate

The Microwave Debate

Nicholas H. Steneck

The MIT Press
Cambridge, Massachusetts
London, England

This book was set in Mergenthaler 202 Baskerville by Achorn Graphic Services, Inc., and printed and bound by Halliday Lithograph in the United States of America

Library of Congress Cataloging in Publication Data

Steneck, Nicholas H.
The microwave debate.

Bibliography: p.
Includes index.
1. Microwaves—Hygienic aspects. 2. Microwaves—Physiological effect. 3. Microwaves—Hygienic aspects—History. 4. Microwaves—Safety regulations—United States. I. Title.
RA569.3.S74 1984 363.1′89 84–7922
ISBN 0–262–19230–6

To Alec and Nicholas

Contents

4

5

II

6

7

8

Preface

Two decades ago microwaves and radio-frequency radiation were seldom topics of heated discussion. Radio, television, radar, and communication equipment were accepted manifestations of an advancing technological society. Today it is often difficult to say anything about microwaves and radio-frequency radiation without being caught in an ongoing debate between a health-minded public and a development-conscious industry.

The story of the development of the microwave debate, the continuing controversy over the safety and use of both microwaves and radio-frequency radiation, is of itself interesting. Its cast of characters includes citizen groups, crusading heroes, international spies, maverick investigators, and media specialists, in addition to the expected array of scientists and policymakers. Its episodes range from routine laboratory discoveries and tedious congressional hearings to delicate international negotiations and classified military research. But there is more to the microwave debate than simply an interesting story.

The problems society faces in learning to use and control RF* energy are endemic to modern technological society as a whole. Accordingly the central issues discussed in this study—public skepticism, controversial media involvement, the bureaucratic structure, conflicting values, vested interests, and scientific uncertainty—are issues that have relevance beyond the bounds of the microwave debate itself. The story told in this

*The abbreviation *RF,* used throughout this book, refers in general to microwave and radiowave electromagnetic radiation. For a more detailed discussion of this and other terms similarly marked with an asterisk, see the technical note that follows this preface.

book is essentially a variation on a theme, not a theme unto itself.

There are advantages to be gained from studying variations to learn about broader themes, which seldom comprise manageable topics for comprehensive analyses. The RF field is small enough to allow a researcher to know and talk with most of the major characters and many minor ones as well. Its literature is large but manageable. The events that lie at the heart of the story are many but not overwhelming. It is possible in this field to see the forest through the trees in a way that larger controversies, such as the current debate over nuclear power or the control of chemical pollutants, defy by their magnitude alone.

Manageable size is a relative measure, however. In recent years episodes of public involvement in the microwave debate have multiplied from a handful to dozens. The professional establishment that daily grapples with the microwave debate has grown from several dozen to several hundred. The technical reports that appear yearly now number in the hundreds, pushing the cumulative literature on which decision making must rest well into the thousands. Size is becoming an important factor to consider. But it is not primarily size that complicates the telling of the microwave story.

Although one of the major characteristics of our age is its propensity for burying the details of daily life in mounds of paper, we are not creating a rich historical record. The information that is routinely filed away in journals, books, and bureaucratic archives is relatively lifeless and uninformative. It fossilizes facts, not the human record that lies behind the facts. We have no systematic way of preserving telephone conversations. We routinely spend billions of dollars abstracting, indexing, and storing technical reports and only a few million keeping track of the more voluminous and informative personal side of life. Few people are trained to keep or even appreciate the importance of their own personal records, and most routinely give way to the problems of limited space.

The consequence of this situation is that the archival record available for reconstructing the recent past is incomplete and difficult to use. Few verbatim transcripts exist from meetings called to discuss RF bioeffects. Their place is filled by brief, epitomized minutes that are more than capable of turning maelstroms into millponds. Transcripts of hearings and other quasi-legal proceedings are usually marred by advocacy, thus making them unlikely sources of balanced, unbiased accounts. Most past attempts to chronicle developments in the RF field have been motivated by ideology, rendering them questionable sources of information. Few of the private letters that have been kept discuss other than routine business. Those that do are usually not readily available (sometimes understandably so) for publication. It is, in brief, difficult to discover how our society is developing, even as it is developing.

This situation will be more troublesome in the future than it is now since most of the microwave story can be reconstructed by turning to the oral record. When I had trouble putting together the pieces of a particular episode, I could pick up the telephone and talk with the individuals involved, a luxury not afforded to historians who study earlier periods. But even this source does not fill all gaps or lead to a coherent story that can be told with confidence. Had I accepted verbatim the personal views related to me during many hours of conversation, I would have had as many stories to tell as interviews generously granted. No two people have the same understanding of how the debate has developed. Were this not the case, there would be no microwave debate and no reason for taking a detailed look at how it came about. But unfortunately it is the case, and as a result there is a problem that needs to be understood before it can be solved.

The primary objective of this book is to aid understanding. In meeting this objective, I have not answered most of the specific questions that are central to the microwave debate itself. Readers who are looking for a simple answer to the question, "Is my microwave oven safe?" or "Will a microwave facility proposed for my community create health problems?" will not find one here. I make a few recommendations in the closing chapter for improving the way in which the microwave debate is handled but take no specific stand on the critical hazards and safety issues. The justification for not taking a stand on such issues is threefold: there remain legitimate questions about the adequacy of the technical data available for making judgments about safety, making such judgments is a subjective process that is best carried out by the public and not individuals, and I do not want the general points made in this book to be lost in the discussions that would inevitably follow as to whether my views on safety and hazards were right or wrong.

Technically minded readers may find the lack of specific guidance frustrating. They may ask why I did not set out the facts, balance them one against the other, and report the final tally. If seeking a final resolution were this simple, such an approach might be appropriate; but the problems that have provoked the microwave debate are not simple and do not admit to simple technical solutions. This being the case, much more attention needs to be paid to understanding, forgetting from time to time the goal of immediate or quick solutions.

This book could not have been written without the cooperation of the scientists, administrators, and members of the public who are engaged in the microwave debate. The tolerant acceptance of a historian into their midst not only facilitated my work but made it enjoyable. If some of my generalizations and conclusions seem in turn a strange way to repay debts, I can only reply that I have tried in return to present a balanced view that will hopefully serve constructive ends.

This study was supported by funding from the National Science Foundation and National Endowment for the Humanities under NSF grants 0SS-78-06675 and 0SS-79-21284. The interdisciplinary focus of NSF's Program for Ethics and Values in Science and Technology made it possible for a historian, a physiologist, and a physicist to work together on a project that fell outside the bounds of each of our fields of immediate expertise. Final thanks are owed to Art Vander and Gordi Kane, the physiologist and physicist who worked with me on this project from the beginning and generously (or wisely) allowed the final product to be published under my name. For those who have not spent hours in intense conversation with colleagues in different disciplines discussing problems of mutual interest, I recommend the experience. It has accounted for some of my most enjoyable and productive hours over the past five years.

Technical Note and Abbreviations

Most of the technical terminology used in this book is explained when introduced. For readers who have little prior knowledge of the microwave debate, it may be useful to keep the following terminology and abbreviations in mind.

ANSI and IEEE Two major nongovernment organizations have played key roles in the microwave debate: the American National Standards Institute (ANSI) and the Institute of Electrical and Electronics Engineers (IEEE). Both have undergone name changes. ANSI was the American Standards Association from 1928 to 1966 and the United States of America Standards Institute from 1966 to 1969; IEEE was called the American Institute of Electrical Engineers prior to 1964. To avoid confusion, I have used the current names when referring to these organizations.

Electromagnetic radiation (EMR) Microwaves are a specific form of EMR. We most commonly encounter EMR as visible light. Less familiar and also less prevalent forms of EMR include radiowaves, ultraviolet and infrared light, X-rays, and cosmic rays. The common element for all of these forms is that they are propagated through space as energy waves that have both electric and magnetic properties; hence the name *electromagnetic radiation.*

Hertz The most common unit used to distinguish different forms of EMR is not wavelength but frequency, the number of waves per unit of time, measured in hertz (Hz; 1 hertz equals a frequency of 1 cycle per second). Frequency is inversely proportional to wavelength, which means that the shorter the wavelength, the higher the frequency. Because frequency differences are large, prefixes are usually used to express increases by a factor of 1000; hence, 1000 hertz equal 1 kilohertz,

1000 kilohertz equal 1 megahertz, and 1000 megahertz equal 1 gigahertz. The radiation associated with electric power lines is usually expressed in hertz, television and radiowaves in kilohertz and megahertz, and microwaves in gigahertz.

Ionizing EMR Ionizing EMR encompasses wavelengths that are shorter than visible light. They possess sufficient energy to change the electronic structure of atoms, producing what are known as ions; hence the designation *ionizing EMR*. X-rays are a form of ionizing EMR. Their high energy and capacity to change the electronic structure of atoms is known to pose definite dangers to health and can cause, among other problems, cancer.

Ionizing versus nonionizing EMR The principal feature used to distinguish forms of EMR one from another is their respective wavelengths. In the visible spectrum different wavelengths are manifest as different colors. Over the full EMR spectrum different wavelengths exhibit another important characteristic, different energy potentials. EM waves of shorter lengths pack more energy than do EM waves of longer lengths. This characteristic is used to separate all EMR into two major subdivisions: ionizing and nonionizing EMR.

mW/cm^2 Just as light bulbs that emit the same frequency (same color) radiation can vary in intensity (for example, 25 watt and 125 watt), nonionizing EMR can be transmitted at different intensities. The unit most commonly used to express the intensity of a particular source is milliwatts per centimeter squared (mW/cm^2), a measure of the power (in milliwatts) per unit of area (square centimeter). Prefixes are used to indicate variations by a factor of 1000; thus 1000 μ(micro)W/cm^2 equal 1 mW/cm^2 and 1000 mW/cm^2 equal 1 W/cm^2. For purposes of general discussion, I have assumed that intensities that fall into the μW/cm^2 range are low level, that intensities between 1 and 10 mW/cm^2 are moderate level, and that intensities above 100 mW/cm^2 are high level.

Nonionizing EMR Nonionizing EMR encompasses wavelengths that are longer than those of visible light. As such, they do not have the energy to change electronic structure and produce ions; hence the designation *nonionizing EMR*. The spectrum of nonionizing EMR includes (from shorter to longer) microwaves, television and radiowaves, the waves that are given off by electric power lines, and extremely low frequency (ELF) radiation. Nonionizing EMR can add energy and heat to matter, including living tissue. Scientists do not agree whether it can produce other effects (nonthermal effects).

RF This book is primarily about radio-frequency and microwave radiation. Rather than using both terms throughout this book, I have used the abbreviation *RF* to refer in general to both. More specific terminology is used when required to portray correctly the technical details of particular discussions.

The Microwave Debate

1

The Skeptical Public

Public: Can you guarantee . . . that those children will be safe?

Industry: It is our plan to present testimony to that effect, yes.

Public: I don't want testimony. I want statements.

Government spokeswoman: This witness has stated that he cannot answer that and there will be testimony later on the question of safety.

Public: Well, you better get a guarantee, honey!

—Home Box Office Hearings, Rockaway Township, New Jersey, June 1980

The technology of satellites, moon trips, and interplanetary exploration has had many down-to-earth applications. Satellite hookups enable major sporting events to be viewed simultaneously in millions of homes. Communication satellites make it possible to talk directly with almost anyone, any place in the world, even where telephone lines have not been stretched. Weather forecasting, mineral prospecting, and crop prediction today are routinely carried on from the perspective of the space-based observer.

The technological spinoffs of the space age have created many new business opportunities. Some of the hardware in space that is charting the weather, directing phone calls, and passing along television programs is privately owned. Initially the capital required for investments in space was large, limiting business activity to major corporate ventures. More recently subcontracting agreements and the space shuttle have made it possible for smaller corporations to move into the space age.

One of the smaller industries that has taken advantage of space-age technology is cable television. The use of communication satellites for independent, nationwide programming gets rid of the need for expensive ground communication links. Cable television companies can direct in-home movies or extended sports coverage to their customers by beaming programs through satellites that communicate directly with earth-based receivers thousands of miles distant from one another. An in-home movie taped in New York can be viewed simultaneously in Little Rock, Arkansas, and Portland, Oregon, by passing it through a single satellite link rather than hundreds of ground communication cables or point-to-point hookups.

In 1975, Home Box Office, Inc. (HBO), a cable television company located in New York City, began satellite programming, a risky move because the highly speculative cable television industry was just beginning to explore the potential of satellite programming. The capital outlay required to implement such a decision could be substantial. Construction costs for an uplink facility, the most visible feature of which is the large dish-like structure that beams the programs to a satellite 20,000 miles in space, could run as high as $100,000. Uplink facilities could be rented, however. RCA had an appropriate transmitter outside New York City near the bottom of a ski run in Vernon Valley, New Jersey, that met HBO's needs. Within a year a rental agreement had been signed, and HBO began beaming nationwide.

HBO's decision to enter the space age proved sound. By 1977, five years after the company began with a single affiliate in Wilkes-Barre, Pennsylvania, their audience had grown to 1 million viewers. Two years later this figure climbed to 4 million viewers signed on by over 2000 affiliates. The success of satellite programming was so phenomenal that HBO decided to invest more heavily in the new technology and build its own uplink facility. Planning for this new facility was well underway by autumn 1979.

The construction of an uplink posed few technological problems. Over four hundred such facilities were already in operation around the world, the technology was readily available, and HBO's requirements were not unusual. They needed a dish that would get programs to any one of a number of appropriate satellites. Purchasing an uplink therefore became a matter of drawing up specifications, receiving bids, and submitting an order. But where to locate the facility was another matter.

Earth-satellite communication links require careful geographic placement. The uplink dish must be able to see all of the satellites it might contact; therefore surrounding topography and existing communication patterns must be such that the uplink signal is not distorted. HBO officials wanted to be close to their New York studio to simplify ground communication and were looking for a pleasant living situation for their personnel. They finally settled on eight or nine criteria necessary to acceptance of a site. In the end only one site reportedly met all the company's needs, a large wooded tract of land located in Rockaway Township in northwest New Jersey.

Confronting the Microwave Problem

Although the targeted site met all selection criteria, it added one major complication. The land HBO wanted to use for industrial purposes was zoned two-acre residential and located in one of the most desirable living areas in the metropolitan area. The site was less than thirty miles from midtown Manhattan and yet within a few minutes of state park lands, a public golf course, and half a dozen lakes. The hills surrounding the site were dotted with the homes of lifelong residents, who had seen enormous growth in the area over the past twenty years, and professional newcomers, many of whom exchanged long daily commutes to work for the relaxation of rural living. Any request to divert a portion of this land to industrial use was bound to run into difficulties.

HBO was well aware of the potential problems when its representatives met with the mayor and town officials in February 1980 to clarify procedures. There seemed little doubt that a request to rezone the entire area would not succeed so a more conservative approach was adopted—application for a land-use variance. HBO was required to submit its plans to a special review panel of the township, the Board of Adjustment, which had the job of evaluating the plans and recommending whether special permission should be given to use the land for industrial purposes. Within a month plans had been drawn up and an application for a variance submitted.

The HBO case should have been simply another zoning case, with residential versus industrial use being the key factor. But it presented a wrinkle: the dishes HBO was planning to erect in the Rockaway woods used microwave radiation to transmit programming to the communication satellites. Any plan to construct a microwave facility in a residential area would raise questions. Even more to the point, the microwave beams had to pass almost directly over the K. D. Malone elementary school, which was located less than a quarter-mile from the proposed transmitters. The citizens of Rockaway had gone through one environmental crisis a few years earlier when industrial wastes threatened to pollute their drinking water. Would HBO and its controversial microwave technology precipitate yet another environmental crisis?

The full extent of local concern over the proposed HBO facility emerged rapidly. Within weeks, groups of objectors began to organize and circulate petitions. When the Board of Adjustment, comprised of eight appointed members, held its first meeting in late March, a preliminary meeting to discuss procedures, the hearing room was packed with a concerned and skeptical public. As it became clear that HBO's application would go forward, a local dentist, Peter Baragona, took steps to block the proposed facility by forming the Concerned Citizens of Rockaway Township. On May 1, 1983, Baragona, the Concerned Citizens, their attorney, and over 100 vocal supporters, some carrying signs and others wearing lapel buttons, crowded into a classroom in the K. D. Malone elementary school to witness the start of the formal airing of HBO's request to move into rural Rockaway Township.

Much of the initial concern over the proposed facility stemmed from opposition to industrial intrusion into a residential setting. Although HBO was hoping to be an ideal neighbor, they could not disguise their industrial features. The proposed headquarters building was designed as a model of corporate architecture, a campus-type structure that blended with the environment and used only about 5 percent of the available land (figure 1). But they had to build as well a 200 foot tower on a neighboring hill and two 11 meter (roughly four story) dishes, which could not be hidden completely from view. On these grounds alone it could be and was questioned whether a community that prided itself on maintaining its rural setting should turn over 50 acres of natural land to industrial use, however passive that use might be. HBO began its case as an unwelcome neighbor at best.

The issue that pushed HBO's status from that of unwelcome to unwanted neighbor was the microwave issue. Had the site

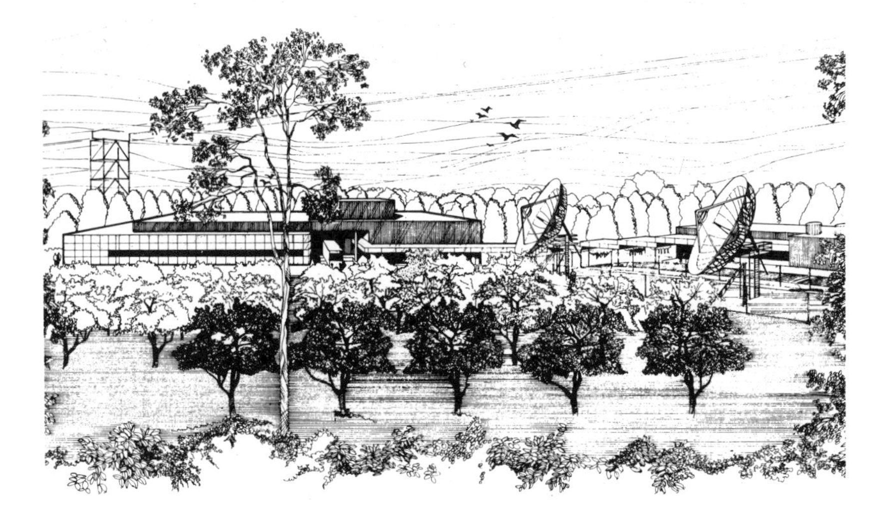

Figure 1
View of the Home Box Office facility proposed for Rockaway Township, submitted to the township in the request for variance.

not been located close to an elementary school, local concern might not have been as great. But it was, and the K. D. Malone school soon became a focal point for the debate. HBO contended that an intervening hill, the distance from the transmitters to the school, and other contingencies, such as a dense tree cover, placed the radiation at the school at such low levels that no hazards existed. The public felt otherwise: "We have read and heard varied information concerning microwave radiation. We are afraid of it. We and our neighbors and fellow parents have a right to protest it. The amount of taxes HBO would pay, whether high or low, does not matter when the health of our children is affected."[1] It remained for the Board of Adjustment to decide whether low-level exposure to microwave radiation posed a threat to public health.

The burden of proof fell on the applicant. HBO was asking Rockaway Township for a special exemption from compliance with one of its laws. Before such an exemption could be granted, HBO officials had to satisfy two objections: they had to prove that the Rockaway site was unique and that the intended use of the land was not detrimental to the public good. In other words HBO had to prove that exposure to microwaves at the proposed levels presented no significant threat to public health.

The person who took on the burden of proof for HBO was a zoning specialist from Morristown, New Jersey, Herb Vogel. In his long career as a zoning lawyer, Vogel had seen many difficult cases successfully through the hearing process. He knew zoning law and how to apply it, but he did not know much about microwaves before accepting the HBO case. His plan of action therefore became one of simultaneous education and demonstration. He planned to educate the public and the board, as indeed he had educated himself, on the technical aspects of the RF problem, while at the same time demonstrating that the proposed facility was carefully planned and safe.

Vogel began his presentation with testimony from HBO's vice-president, Glen Britt, who laid out the corporation's history, plans, and concerns. Britt was to be followed by the building's architect, who would describe the facility. Thereafter Vogel intended to call a communication specialist, who would present the technical details of the system and provide expert testimony on a major concern of everyone involved, actual exposure levels. Then one or more experts on the biological effects of microwaves were slated to testify on the degree to which harmful effects would be expected at the levels of radiation present. Assuming that the bioeffects experts agreed that no hazards would be expected at the projected levels of exposure, Vogel was confident that he could overcome local resistance and be granted the land-use variance.

Problems began to arise from the start. Britt had been instrumental in making many of the decisions that led to the selection of the Rockaway site, but he was not qualified to testify on most of the issues that had to be discussed before the variance could be granted. He was not an engineer so could not describe the safety features of the proposed system. He was not a communication expert so could not comment on the details of the site selection process. He had not participated in the research on consumer demands so could not testify on the specifics of HBO's projected corporate growth. Again and again he was compelled under cross-examination to ask for patience.

But patience was not something that an anxious public was long on. Too many details of the application did not make sense, and the public wanted immediate explanations. Why was the microwave beam aimed south over the school when Britt's office and the building's windows were all on the north, facing away from the dishes? Why didn't Britt intend to relocate his own family to Rockaway? What was the company thinking when it planned to cut down the trees in the path of the beam because they were too tall and replace them with apple trees? Didn't they realize that the apple trees would draw children to HBO's property and directly into the path of the microwave radiation? And what about hikers? Would they be kept away by fences? If fences were built, how would firefighters gain access to the facilities? If firefighters did gain access to the facility, would they be warned of dangers? Would there be a danger that the 200 foot tower, which was located less than 200 feet from the property line, might fall on persons in the area? And most important, "Would Home Box Office be willing to put in writing their guarantee that there will be no harmful effects to anyone or anything outside their property line?"[2]

Faced with this barrage of questions and especially the request for a guarantee, Britt provided the most logical reply under the circumstances: "I'm not sure that it's possible to do that [guarantee safety] with anything in life."[3] But uncertainty was not something that a skeptical public was likely to accept. It was, after all, the public's children and not HBO who had to live with the consequences of a decision based on uncertainty.

In demanding guarantees the public was asking for a specific statement about safe and hazardous exposure levels. The basis for such statements is usually a standard, a formally established reference point that can be used for making judgments about other related points. Using existing standards as his guide, Britt initially offered the following guarantee: "We are going to present evidence that we [will] meet every existing standard for [microwave radiation], the toughest standard in the world."[4] In other words, to allay public fears, HBO agreed not to expose anyone to levels of radiation that had been judged to be dangerous by any recognized expert in the world. Surely this should be an acceptable guarantee once proved. But was it?

Standards can be trusted only if the experts who establish them are trusted. They are no more reliable than the government, industrial, and academic communities that assemble and weigh the scientific evidence on which they rest. And it was precisely these experts and the evidence they assembled that many citizens of Rockaway Township distrusted. As one woman candidly put the public point of view, "All the evidence doesn't amount to a piddle hole in the snow."[5]

Many members of the public had serious reservations about

the information on the effects of low-level exposure that had been used to reach conclusions about RF hazards. Before the hearings began, the executive board of the local PTA voted unanimously "to oppose construction . . . based upon *inconclusive evidence* as to potential harm caused by such microwave installations."[6] Their action was actively supported by roughly two of every three families affected by HBO's plans. A few weeks later, still well before any experts had testified on bioeffects, the Board of Education had "reviewed all of the information and data that is available" and concluded unanimously "that there is reasonable doubt as to the safety of that installation."[7] The members of the public were not willing to rest their judgment on traditional authorities; they wanted HBO to take personal responsibility for the facility and its consequences.

Britt finally accepted this responsibility late on his third night of testimony. (Hearings, normally held one night a month, frequently pushed on to near midnight.) He again reiterated HBO's promise to comply with all present standards, adding that HBO would "expeditiously" conform to any future standards of safety requirements. If compliance with standards were not sufficient, then HBO was "willing to and does guarantee to the Rockaway Township Board of Adjustment that no adverse health effects will result to any citizen or citizens of Rockaway Township from the operation of the proposed microwave facility."[8]

Even this unprecedented guarantee did not get HBO off the hook. It promised little more than that sometime in the future the public could go to court for damages if injury could be proved. This represented at best an expensive and uncertain way of being paid for damages that the public did not want to risk incurring at all. The guarantee faired no better than the evidence: "The guarantee doesn't mean a piddle hole in the snow. Put that on the record."[9]

Breakdown of Dialogue

Given the public's initial skepticism and hostility, Vogel's task was to rebuild confidence on the basis of expert opinion. This crucial point in the hearing process was reached on the eighth night of testimony. After two months and more than twenty hours devoted to background information, the first expert took the stand on July 10, 1980.

Experts, like standards, have credibility problems. Their reliability is not absolute. The language they are called on to interpret, the language of science and technology, is not free from ambiguities. Facts, including many scientific facts, do not speak for themselves. In cases of ambiguity the expert as interpreter has to supply not only a translation but a meaning, a process that requires judgments about the validity of research methods, the significance of experimental data, and so on. Although the main reason for calling on experts is to clarify the facts, the testimony they ultimately provide is opinion testimony based on their own personal experience.

Because the testimony of experts rests in part on personal opinion, the background experts bring with them into the hearing room is important. When an expert testifies that no significant adverse health effects were seen in a study of workers exposed to microwaves, that expert's definition of *significant* is critical. Is injury to a single worker significant or only a statistical increase in adverse effects? The public, which has to weigh expert opinion and decide on its relevance, has a vested interest in screening experts carefully before listening to their testimony.

The wrangle over experts in the HBO hearings took many twists. One potential witness, Stephen Cleary of the University of Virginia, was recognized by all concerned as a person with ample expertise. While a young researcher in the late 1960s, he had played a major role in the planning sessions used to hammer out RF bioeffects research policy. His own research made important contributions to that policy, as did his presence on numerous committees. There was no doubt that Cleary was a qualified expert. But who would call him?

HBO officials made the first contacts. They talked with Cleary by telephone and then paid him a consultant's fee to come to New York and discuss testifying. But after this initial contact, they had some reservations. They were entering into an adversarial proceeding in which they had not only to prove their case but to prove it in the most effective manner possible. They trusted Cleary's judgment but were uncertain about his impact in the hearing room.

While HBO was pondering this problem, the board's attorney, Lee Greb, informed Vogel that he too was considering calling Cleary. This raised a further complication. Would the fact that HBO had contacted Cleary first and paid him prejudice his testimony? Would anyone believe him if, for example, he were called by the opposition but agreed with HBO's position? Vogel had his doubts. Moreover, since under Board of Adjustment procedures HBO had to pay the expenses of experts called by the board, he wanted to make sure that the experts called were believed. Accordingly Vogel objected to Cleary's being used by the board. The board in turn objected to HBO's potentially discrediting witnesses by retaining their services for a day or two.

Cleary's status had not been resolved by the eighth hearing when the first expert, James Cuddihy, was sworn in. Cuddihy's role as chief engineer and consultant for the proposed facility made him the logical witness to call to testify on technical details. He was familiar with microwave engineering and thus could provide an expert opinion on one of the most crucial and discussed pieces of information in the entire application process, the levels of microwave radiation that would be present at the K. D. Malone school if the proposed transmitters were built and turned on.

Cuddihy was well qualified to testify on these topics. Prior to working for HBO, he had spent twelve-and-one-half years with RCA working with every phase of its satellite communication program. He had designed components, developed whole systems, overseen construction, and followed transmitters through the licensing process to successful operation. He had worked on nearly 40 of the some 400 transmitters that had come into operation by the summer of 1980. His experience ranged from the Vernon Valley transmitter HBO currently was using to a facility in China hurriedly put together at the request of the State Department to cover President Nixon's trip to Shanghai. But although Cuddihy's experience seemed ample, he lacked one formal qualification that initially caused some problems; he was not a licensed engineer in the state of New Jersey. Could he then legally help design a facility that was to operate in New Jersey and, more to the point, testify about its operation? After close to an hour of legal discussion, his testimony was provisionally accepted with the stipulation that it might be disqualified sometime in the future.

Once the legal bridges were crossed, the public and board got down to the business of testing Cuddihy as a scientist and engineer. Questions were raised about the formulas he had used to estimate field strengths, the degree to which these formulas were accepted by the scientific community, and even the reliability of the textbooks in which they were published. HBO was confident of Cuddihy's expertise. His examiners were less confident. They did not believe in the old adage that figures do not lie. The clear implication of a great deal of the questioning by both board and public was that figures and formulas can be manipulated and that mathematics cannot necessarily be trusted.

Cuddihy's estimates of exposure levels did not rest solely on mathematical calculations. He realized that a lay public might not accept calculations they could not understand and so convinced HBO to pay for field measurements at a transmitter similar to the one that was to be built. The logical choice was the Vernon Valley transmitter, which was about the same size, of comparable power capacity, and situated so that the geographic location of the school could be fairly closely duplicated.

The proposed measurements at the Vernon Valley site were made shortly before the hearings began. Special instruments were flown from Texas to Newark Airport. The specialists who accompanied the instruments from Texas drove them to Vernon Valley, where, under Cuddihy's supervision, the measurements were taken. Cuddihy personally visited the facility at the bottom of the ski slope and checked with the RCA engineers in charge to make sure that the transmitters were operating when the readings were taken. He photographed the meters and made his own initial calculations so that there could be no doubt about the values being recorded. Armed with this seemingly concrete information, he confidently testified that the radiation levels at the K. D. Malone school would be thousands of times below the most rigid RF standard in the world and a million times below the U.S. standard.

As solid as Cuddihy's testimony was on actual exposure levels, it was not above questioning. Although similar, the Vernon Valley and proposed sites were not the same. The differences could be crucial. Then there was the matter of the instruments used to make the measurements. They had been flown by commercial airliner from Texas to Newark and then driven by car to the site. At the site they had not been checked for accuracy. Possibly the delicate instruments had been damaged in transit and were not reliable. Members of the board even questioned Cuddihy's ability to conduct field tests. His experience lay in the area of system design, not field measurements. Could he be relied on to provide expert testimony on field radiation levels?

Before Cuddihy could end his testimony, the biological expert, Herman Schwan, had to be squeezed in to accommodate an upcoming leave in Germany. In most respects Schwan's credentials were even more impressive than Cuddihy's. A native German he had begun work on the biological effects of electromagnetic radiation* before microwaves (as radar) were developed at the beginning of World War II. After the war he continued his work in the United States, becoming a key figure in many research and policy decisions over the next thirty years. In 1953 he formulated the well-known 10mW/cm^2 (10 milliwatts per centimeter squared)* figure for regulating exposure to RF energy. Slightly more than a decade later he was instrumental in having this figure formally adopted as the American National Standards Institute's official RF standard. He had organized international symposia, testified before Congress, and authored or coauthored hundreds of scientific papers. Perhaps most important he possessed an unusual capacity to explain scientific information in a clear, convincing, and, usually, reassuring manner.

As broad as Schwan's qualifications were, like all other modern scientists he was a specialist, a fact that formed the opening wedge for questions about his qualifications. His particular specialty was biophysics; he applied physics to biology. Thus Schwan had not worked extensively with actual biological materials and particularly with human tissues, but he was there to testify about what might happen to children exposed to microwaves. Could he be accepted as an expert witness if he were not testifying about his own experience in the laboratory? This question was raised repeatedly over Schwan's five evenings of testimony, which stretched from mid-July well into August.

Question: I think that I had asked you on a prior occasion whether or not you have examined human organs in regard to whether they had been affected by microwaves that had been projected onto those human organs.

Schwan: No we haven't.

Question: Then the bulk of your testimony is based upon an estimate and not any actual experience with human tissue?

Schwan: That's completely wrong! That's entirely wrong! We have discussed that for hours. . . .

Chairman (intervening a few moments later): If I understand the question correctly, it is simply have you personally done studies on human tissue that have been exposed, living human tissue?

Schwan: My answer was no.

Chairman: Okay. And basically what he said is that your findings then are the result of inference from studies, your contacts with colleagues, but not personal experience?

Schwan: Mr. Chairman, what he is after [is] how can I be judged in this field if I haven't done studies myself with human tissues that have been exposed. That's the implication. That's what he is driving at. And I have answered that question.

If you use his procedure you can discredit any witness on the stand, since you can then discredit the one who has done medical experiments, but has no biophysical insights and no bioengineering insights. How can you make the statement that the standard should be formulated so-and-so if you don't know how to extrapolate from animal experiments to man?

You can kill any witness and declare that he is not a witness of standing.[10]

Inability to provide general testimony on the basis of personal experience was not the only obstacle that threatened to undermine Schwan's expert testimony. Although Schwan had agreed to testify for HBO, as a scientist he could not fully concur with its position. Britt had said that HBO would guarantee no adverse health effects, but Schwan refused to give an actual guarantee: "I cannot prove the impossible. I cannot prove that subtle effects of microwaves could not exist for sure."[11] He regarded the possibility of subtle, unforeseen effects as extremely unlikely but could not say that there would never be any effects. He could not, in sum, provide certain evidence to resolve the issue of RF bioeffects.

There was yet another obstacle that threatened to undermine Schwan's expert testimony. His long association with the RF bioeffects field could be seen as enhancing his credentials, given the extent of his experience, or as evidence of associations that would make it extremely difficult for him to break with prevailing points of view. The question of Schwan's loyalties came up during public questioning late on his third night of testimony.

Question: May I ask you about C95.IV, which is the committee that you chaired at one time? It was reported by Paul Brodeur that C95.IV felt obliged to protect the 10 milliwatt level at all costs and to ignore, deny or, [worse] come to worst, suppress any information about adverse effects of low intensity microwave radiation. Would you care to comment about that?

Schwan: It's an unsubstantiated lie![12]

Two sessions later, again during the closing hour of debate when the proceedings frequently heated up, the same person stepped up to ask a question. With the prior experience in mind, Schwan refused to respond orally to the question, saying he would provide written answers. Objections were raised and several minutes of confused debate followed. Finally the chair ruled that the questioning should go forward and that Schwan should answer. Rather than answer, Schwan stood up, bid the chair goodbye, and left the hearing room.

Although two more nights of hearings followed Schwan's testimony, it became increasingly clear to Vogel and HBO officials that a long and uncertain road lay ahead. Between May 1 and October 9 seventeen hearings had been held. Four witnesses had been called, none of whom had finished testifying. Conservative estimates placed the number of future hearings necessary to complete the case at thirty, assuming that the list of witnesses for the board and public was not extensive. At one or two meetings a month this placed the final decision well over a year and many tens of thousands of dollars away. The price in terms of dollars and time was too high. HBO withdrew its application.

Scope of Public Concern

Shortly after the HBO hearings ground to a halt in the fall of 1980, a COMSAT (Communications Satellite Corporation) project proposed for Columbia County in eastern Pennsylvania (surrounding Bloomsburg) ran into similar opposition. This time economic development held sway over the objections of a dozen or so citizens, and the project went ahead. As fears over the safety of the proposed COMSAT facility were being quelled, a new pocket of concern emerged on the West Coast on Bainbridge Island, an unspoiled rural area near Seattle, Washington. A few years later a request to add an additional uplink dish at the original Vernon Valley site spawned yet another debate. On many occasions and in many different settings the public has expressed deep reservations about the desirability of satellite communication technology.

As important as the television uplink cases have been, they represent only a small fraction of the total public energy devoted to questioning the increased presence of RF energy in the environment. Over the past decade nearly every form of RF technology has been the target of public concern at one time or another. Radio broadcast towers, microwave relay systems, video display terminals, radio-frequency heat sealers, radar systems, electric power lines, the U.S. Navy's ELF/SANGUINE/ Seafarer submarine communication grid, and the Solar Power Satellite project have each had their day before the public. According to one observer the RF problem has become "the biggest environmental hot potato" of the day.[13]

One common RF product that has been on the public mind since its widespread marketing began in the 1960s is the microwave oven. By the late 1960s reports of faulty ovens were becoming a commonplace in the press. Consumers Union ran tests in 1973 and promptly stamped on its discussion of microwave ovens "not recommended" because it could not "uncover any data establishing to our satisfaction what level of microwave radiation emission can unequivocally be called safe."[14] Therefore CU could not in good conscience recommend them to the potentially millions of consumers who relied on their advice.

CU has softened its stand since 1973. A later report on microwave ovens in March 1981, while stopping short of endorsement on safety, assured the public that "microwave ovens can be operated with a minimum of risk from radiation" if some fairly basic precautions are taken. This is not to say that there are no risks. CU still cannot find "persuasive evidence" to suggest that "exposure is completely without risk." But the risks, if there are any, now seem to fall into an acceptable range.[15]

Radar systems became particularly troublesome when the U.S. Air Force proposed to increase the power levels of some of its coastal defense units by installing a new system, PAVE PAWS (Precision Acquisition of Vehicle Entry Phased Array Warning System). Two units planned for the West and East Coasts, respectively, ran into public opposition. In the sparsely populated area surrounding Beale Air Force Base in California, a loosely knit coalition, Citizens Concerned about PAVE PAWS, brought a National Environmental Policy Act suit to attempt to stop construction. In the more heavily populated Cape Cod area surrounding Otis Air Force Base, the Outer Cape Environmental Association, the Cape Cod Citizens concerned about PAVE PAWS, the Truro Citizens Concerned about PAVE PAWS, and several less formal groups spent hours searching for a responsible environmental policy. Both radar projects were finally completed despite local opposition and are now operational, but the environmental issues that surround them have not been forgotten.

Public concern over the PAVE PAWS system has focused on the present and future. A few years earlier another radar-oriented group emerged that was troubled about the past. In the late 1960s a retired air force radarman, Joe Towne, discovered that he was slowly losing his sight as a result of cataracts. Through his own informal surveys and an official government study, Towne came to suspect that the cause of his eye problems might have been the RF exposure he suffered during his ten years of active duty on EC 121 surveillance planes. When official response to his charges proved unsatisfactory, Towne began a one-man campaign to publicize the plight of possible radar victims.

By July 1976 the response to Towne's campaign was sufficient to allow him to call the first meeting of the Radar Victims Network in Framingham, Massachusetts. Twenty men made the first meeting, at their own personal expense and, in some cases, despite serious medical problems, ranging from blindness and advanced cases of cancer to heart problems and blood disorders. Their purpose in organizing was to "help others who have also suffered damage to health from their exposure to electromagnetic radiation."[16] Within a year Towne's organization had a mailing list of 175 names, many of whom were fighting cases or who had won some form of compensation for their injuries.

Concern over radar brought organized labor into the microwave debate by the mid-1970s. In October 1976 stories about radar hazards circulating in the popular press prompted the president of the National Association of Government Employees, Kenneth Lyons, to write to the secretary of transportation, William Coleman, asking for an immediate investigation. To put some clout behind his request, Lyons noted, "Our organization is preparing to negotiate a nationwide contract in behalf of FAA technicians, but until we get the information concerning the above [a number of specific problems], I doubt very much that we could properly negotiate a meaningful contract."[17] Coleman replied, assuring Lyon that "the safeguards established by the FAA afford maximum protection from radiation." FAA's radar workers were not, and still are not, convinced. Their official organization, FASTA (Federal Aviation Science and Technological Association), still believes that "the problem of FAA technicians being exposed . . . to dangerous levels of radiation is a constant concern" and is keeping a wary eye on the expanding use of RF technology.[18]

From radar, organized labor turned its attention to industrial RF technology, particularly video display terminals (VDTs) and RF sealers. VDTs are essentially television sets adapted to industrial applications. Most of the country's newspapers are now composed with computers, which display the content of their programs (tomorrow's newspaper) on television screens. Newspaper composers therefore sit in front of a television screen or VDT most of the working day. Stories linking these exposure conditions to cataracts brought the Newspaper Guild of New York into the microwave debate in late 1976 when suits were filed against the *New York Times* alleging injuries to two workers. The suits were eventually dismissed, and a subsequent agreement to seek an opinion through arbitration turned up no specific correlations between VDTs and health effects.[19] But the organizations representing employees continued to monitor the situation.

RF sealers have recently aroused widespread labor interest. RF energy has the ability to heat select materials rapidly and economically. As is well known, a turkey can be cooked in a microwave oven in less time than in a conventional oven. Similarly glues can be dried rapidly by placing them in an RF field, and plastics and other materials can be cooked, bonded, sealed, and otherwise manipulated with RF sealers and heaters.

An estimated 1 million workers are exposed to RF energy through such devices, many for major portions of the working day, at levels that approach or even exceed current safety guidelines and during susceptible periods in life. A primary concern are expectant mothers, who sometimes sit near RF sealers during pregnancy for eight hours a day. One congressman who recently learned these facts during hearings before a subcommittee of the House Committee on Science and Technology was "flabbergasted" and "frankly shocked" by what he heard.[20] The AFL-CIO is examining the exposure statistics before taking further action. Clearly RF sealers and heaters could replace radar and uplink facilities in the 1980s as a primary rallying point for public interest and labor groups.[21]

It is unlikely that the microwave debate will disappear over the next few years. The uses of RF technology are multiplying more rapidly than the ability to test and regulate them. Since the late 1970s the pace of regulation has slackened in the United States, perhaps leaving more and not fewer problems with which to deal. An alternate route for addressing the RF problem, the legal route, is becoming clogged with litigation. Economic and political pressures seem to leave no room for retreat from increased dependence on RF technology. Social concerns seem to leave less and less room for the deployment of that technology in the public sector. As these forces converge, an inevitable clash of technology, science, and values seems to loom ever nearer on the horizon. That such a clash will occur (or is already occurring) is not in doubt. The only question that remains is how serious its consequences will be—for both the producers and user of RF technology and the communities that surround them.

In this book I have used the tools of history to bring order to our understanding of the past and ongoing development of the microwave debate. The opening chapters (2–5) cover the private years, the years when debate and decision making were carried on mostly within the community of experts who were exploiting RF technology for military and industrial purposes. Chapter 2 discusses how the experts dealt with the RF problem when it first emerged in the 1930s and as it continued to haunt them during the 1940s, 1950s, and 1960s. The major decisions about safe exposure levels, setting standards, are covered in chapter 3. Chapter 4 delves into the problems created by data and ways of thinking that did not fit with the prevailing thermal point of view, problems that took on international significance when the United States discovered in the early 1960s that the Soviets were bombarding the U.S. embassy in Moscow with weak microwave radiation. The steps the U.S. government took to deal with the "Moscow signal" are covered in the concluding chapter on the private years.

The public years began in the late 1960s when the federal government extended its environmentalist concerns to the radiation problem. Chapter 6 briefly discusses the efforts government has undertaken to solve the RF problem. Chapters 7 through 9 turn to three other arenas in which the microwave debate has been carried on and is currently raging: the scientific community, the media, and the courts. The accomplishments and failures of each of these institutions are discussed in turn, thereby bringing the microwave debate to its present state. The final step, from present to future, is left for the concluding chapter.

I

THE PRIVATE YEARS, 1930–1967

2

The Thermal Solution

An examination of actual scientific results dealing with biological effects of microwave radiation indicates that adverse biological effects resulting from such radiation occur largely as a result of thermal injury of biological tissues incurred at power densities in excess of $10 mW/cm^2$. . . . Unlike ionizing radiation, microwave radiation does not appear to exert a cumulative effect with respect to tissue damage. This is true even with repetitive exposure to radiation at power levels sufficiently high to cause thermal injury, if recovery of the injured tissue is permitted to occur between exposures. Moreover, there is no evidence of an induction or latency period for microwave radiation with respect to chronic disease processes.—"Batelle Human Affairs Research Center, Final Environmental Impact Statement, RCA Earth Station, Bainbridge Island, Washington, June 1982

Concerned citizens acting together to protect themselves from perceived threats to their well-being became a common feature of American life in the 1970s. The experience of Love Canal, Three Mile Island, and other environmental crises made the public wary of anything new and scientific or technological. Microwaves, the product of recent scientific and technological development, readily fell within this suspected triumvirate.

RF technology did not grow up and mature in an atmosphere of suspicion. When the newly invented ultra-high-frequency radiowaves (later called microwaves) were first brought to the attention of the public in 1942 and 1943, they could not have been more welcomed. At the time the United States and its allies were engaged in a bitter war whose outcome seemed heavily dependent on the abilities of the opposing powers to put science and technology to work. Inventions that were new, scientific, and technological and that worked were looked at as the preservers, not the annihilators, of peace.

Microwaves occupied a premier place in the new arsenal of scientific weapons introduced during World War II. They could be used to detect and locate enemy aircraft, ships, factories, and other instruments of war in any weather and at any time of day or night. Radar installations along the coast of Britain helped the English withstand the Luftwaffe raids of autumn 1940. During the opening skirmishes in the Solomon Islands, they permitted a U.S. ship to locate and destroy a Japanese vessel so distant that it was never actually seen (although the flames of the burning ship were visible once it had been hit with radar-directed naval gunfire). By mid-1943, when newspapers around the world began to carry stories on the new wartime technology, it was under such headings as "Radar—Our Miracle Ally." Americans counted on radar technology to help win the war, and they looked beyond "when this new branch of science can be turned toward the purposes of peace." The public's first impressions of microwave technology were conditioned by the belief that it could "perform a multitude of tasks that will relieve men from drudgery and bring higher standards of living and culture."[1]

This attitude toward RF technology did not change significantly until the late 1960s. Radar was useful not only in war but in peacetime. It aided commercial aviation and improved weather forecasting. Microwave technology led to advances in the communications industry and had potential for revolutionizing domestic cooking with the invention of the microwave oven. Improved radio-frequency technology brought with it the age of colored television. During the immediate postwar era, when progress was looked at as "our most important product," a General Electric advertising slogan of the 1950s, RF technology contributed its share to progress and reaped appropriate economic and social rewards.

The importance of the early years of overwhelming and enthusiastic support for RF technology in the current microwave debate cannot be overemphasized. The microwave debate was begun during the progress-oriented pre-Vietnam, pre-Watergate era, roughly before 1965, and its ground rules were established well in advance of the heightened environmentalist mentality of the late 1960s and early 1970s. This fact is indispensable for understanding the historical developments discussed in this book. The patterns of thought established in the 1930s and 1940s continued to exercise a major influence on the development of the RF field well into the 1970s and even into the 1980s. Accordingly it is necessary to begin with a discussion of these early patterns of thought.

Radiowaves and Bioeffects

There has never been any serious doubt about the causal relationship between RF energy and bioeffects. Shortly after the development of radiowave technology in the late nineteenth century, medical researchers discovered that radio-frequency electric current can affect living tissue. Within a few years medical practitioners were using radio-frequency electricity to treat many ailments. As radio-frequency radiation (radiowave) technology developed during the early 1900s, it too was exploited in medical treatment. By the 1930s diathermy, as radio-frequency treatment came to be known, was accepted as a beneficial new technology by many physicians and was being used to treat everything from backaches and muscle pain to cancer.[2]

There has also never been any serious doubt about the dangers posed by excessive exposure to RF energy. As early as 1928 a researcher at the Albany Medical College, Helen Hosmer, showed that radiowaves can heat body tissue. Hosmer had been called to nearby Schenectady, New York, to investigate the effects of experimental short-wave radio transmitters on workers at a General Electric research facility. An examination of one person working near a transmitter showed that his temperature rose 2.2°F after only fifteen minutes of exposure. If brief exposure could produce a modest amount of heating, presumably more extended exposure could produce more extreme heating. The conclusion that needed to be drawn was obvious. Although tentatively suggesting that radiowaves might be useful in therapy, Hosmer went on to caution that "it would seem advisable to use extreme care and to postpone until after thorough investigation the exposure of human beings to the more powerful types of this [radiowave] apparatus. We cannot yet predict with certainty just where extreme local heating might occur before the general body temperature gave sufficient warning."[3]

In the late 1920s, however, there was ample reason to be

more excited than troubled by the discovery of heat effects. Many physicians believed that artificially produced fevers could cure diseases. Fever was obviously associated with the body's natural curing mechanism. It therefore seemed reasonable to conclude that "artificial" fevers might produce "artificial" cures. Many techniques had been developed to produce artificial fevers, some far from satisfactory. One technique required the introduction of fever-producing diseases, such as malaria, into the body to raise body temperature and thereby presumably bring about recovery. Radiowave technology held out the possibility of specific controlled heating without the introduction of potentially dangerous foreign agents, such as other diseases.

By 1930 research on the therapeutic value of radiowave-induced fevers had swung into high gear in the United States and abroad. Over the next decade international conferences were called to discuss radiowave fever therapy, and hundreds of articles appeared discussing the seemingly limitless applications of diathermy treatment. In the minds of many a new era in physical medicine was beginning. One group of researchers at the General Electric plant in Schenectady, where Hosmer's initial observations had been made, developed a box-like apparatus that could increase a patient's body temperature about 1°F for every fifteen minutes of exposure. The patient was laid on a frame criss-crossed with a web of cotton tapes and placed between two aluminum plates. The patient and plates were then surrounded by a cellotex box, which kept heat from escaping. Openings were left at the head and foot for the circulation of air and to add additional heat when required. The aluminum plates were then connected to a radiowave generator. The waves generated between the plates passed through the patient's body and steadily increased body temperature. Readings as high as 106.5°F were recorded, and it seemed plausible to suggest that higher temperatures might be obtained easily with the apparatus.

Since the consequences of higher or prolonged temperatures were not known, the GE team took Hosmer's earlier advice and proceeded cautiously, with good reason, for prolonged high temperature could produce adverse consequences. A few patients studied at GE complained of headaches, felt nauseous, and/or experienced drops in blood pressure. But none of these symptoms was unduly worrisome; they were, after all, the same symptoms normally experienced during illnesses that cause fever. Moreover the patients did "not appear to be greatly distressed or fatigued when the maximum temperature is maintained for one hour and then allowed to return to normal while the patient is well blanketed." If used with caution, selective heating with radiowaves appeared to be a safe way to raise body temperature.[4]

Although the effects associated with controlled heating were not looked on as troublesome by physicians, especially if done for the purpose of therapy, uncontrolled heating outside the medical context was another matter. Heat stress on the job can have serious consequences. Vigorous work combined with high temperatures can cause adverse effects, from general fatigue and increased pulse rate to heat stroke. With this thought in mind the navy's Bureau of Medicine and Surgery in July 1930 began an investigation of the potential hazards posed by the new super-high frequency (80MHz)* radio transmitters being put into operation.

The effects of the new transmitters were quite startling in some cases. A normal incandescent light bulb reportedly glowed when held about twenty feet away from such a transmitter. The bulb glowed because the body of the person holding it acted as a receiving antenna and absorbed enough electricity to cause the bulb's filament to illuminate. Moreover the absorbed RF energy produced an "unpleasant warmth and sweating of the feet and legs, . . . general body warmth and sweating, drowsiness, headaches, pains about the ankles, wrists, and elbows, weakness, and vertigo." The purpose of the navy's test was to find out how severe these symptoms could become and whether they produced any permanent adverse effects.

The navy's test program followed a straightforward investigative format. Six volunteers were asked to stand near a super-high frequency transmitter and endure effects until they became unbearable. To increase the production of effects, each volunteer held a metal rod in one hand and grounded the rod on a metal fence, which ensured maximum absorption of electromagnetic energy. The power was then switched on and the symptoms recorded in order of appearance.

The first symptom experienced was a growing sensation of warmth in the hand holding the metal rod. The heat became intense enough within three to five minutes to cause the volunteers to shift the rod from hand to hand. Next pain began to develop in the joints and tendons of the hand holding the rod.

The pain was not localized and could not be changed by external pressure. Similar pain soon developed in the ankles. General sweating followed, along with a feeling of increased total body temperature. Tests showed that body temperature was in fact being elevated a few degrees and that there were drops in blood pressure. Finally, the feeling of heat was followed by "weakness, drowsiness, or headache."

All of these symptoms disappeared when the transmitter was turned off, and the volunteers seemed to return to normal in all respects except one, the rate at which subsequent exposure produced effects. Volunteers who were subjected to a second and third series of tests experienced all of the symptoms faster than when they first underwent testing and took a longer time to recover from the exposure sessions. All did return to normal, however, at least to the extent demonstrated by the tests conducted. No one at the time was concerned with long-term effects; thus long-term factors were not taken into consideration in looking at possible hazards.[5]

The results of the navy tests for the most part confirmed the observations made by medical researchers. Radiowave-induced heating seemed to produce the same symptoms as fevers: sweating appeared first, followed by lower blood pressure, headaches, nausea, and dizziness. The navy's results also fell into the general pattern of effects observed in workers exposed to high-temperature work conditions. A 1921 study of miners working under high temperature and high humidity conditions listed almost point for point the same symptoms of overheating: low blood pressure, increased body temperature, excessive sweating, dizziness, headaches, and nausea. Further evidence of the relationship between fever symptoms and hot-work environment symptoms were provided by the fact that three of the volunteers the navy tested had themselves worked in high-temperature environments in ship firerooms. Each reported that the symptoms he experienced in both situations were similar.[6]

Given the similarities of radiowave-induced heating, fevers, and work-related heat stress, a clear thermal framework began to emerge over the course of the 1930s for resolving the bioeffects or hazards issue. What dangers to human health did the new radiowave equipment pose? Majority opinion overwhelmingly sided with the conclusion presented at the end of the 1930 navy study: "from a practical point of view there are none."[7]

In denying the existence of dangers, neither the navy nor anyone else was suggesting that exposure to radiowave energy could not produce adverse effects. Radiowave equipment was not being labeled benign. If used injudiciously, the radio, like tanks, airplanes, or any other piece of high-technology equipment, could cause injuries; however, from the point of view of day-to-day experience, the chance of injury did not appear extreme or uncontrollable. If proper precautions were taken—if exposure times were kept to a minimum, protective screening were employed whenever possible, work rooms were adequately ventilated—the worst that could be expected from exposure to high-powered radiowave equipment was some overheating along with the symptoms that accompany fevers. Such consequences did not appear unreasonable given the significant benefits the use of the new equipment promised.

The focus on the tolerability of heating effects comprises the heart of the thermal solution to the radiowave and later microwave bioeffects problem. By the mid-1930s most researchers who studied RF bioeffects were convinced that all significant radiowave bioeffects are thermal in origin and that thermal effects are tolerable and reversible if kept within reasonable limits. Therefore they were not reluctant to develop further radio-frequency technology for the improvement of human life. For all intents and purposes, with the formulation of the thermal solution, the RF bioeffects problem was considered solved.

Wartime Technology and Postwar Problems

The invention of radar early in World War II interjected a new ingredient into the microwave debate. Radar uses short radiowaves, or microwaves. Typically the electromagnetic waves that transmit radio programs are roughly 3 to 300 meters long; microwaves fall into the range of 3 to 30 centimeters. Microwaves pack more energy than radiowaves and thus have the potential for causing greater harm, other factors being equal. Determining how much more harm predictably rekindled the bioeffects debate.

Every time the bioeffects issue was rekindled throughout the war years, thinking invariably came back to the medical sphere and thermal considerations. In 1942 forty-five navy personnel who routinely worked with short-wave radio and radar were subjected to a twelve-months study that included blood tests, physical exams, and case histories. The research team that conducted the study reportedly found no evidence of significant effects in the test results. A few radar workers experienced the now-familiar heat-related symptoms: headaches, warming of the extremities, and a flushed feeling. None of these effects persisted, however, so there appeared to be no cause for concern, given the fact that "the radio-frequency energy of radar" seemed no "different from that of other high-frequency radio or diathermy units of an equivalent average power."[8]

A similar study by the Aero Medical Laboratory in Boca Raton, Florida, in 1945, this time of 124 officers and enlisted men, led to essentially the same conclusion. "No evidence" of altered blood counts was found in "personnel exposed to emanations from standard radar sets over prolonged periods." The negative results were again placed squarely within the medical and thermal contexts. Quoting a National Research Council Report, the U.S. Army Air Force researchers noted that "the power which can be dissipated in a subject exposed to microwaves from the radar systems of highest power under the worst possible conditions . . . is of the same magnitude as that used in high-frequency therapy."[9] Duty at a radar station and a trip to a doctor's office seemed to present essentially the same degree of exposure to radio- and microwave radiation.

Although correlations were frequently made during the war years between diathermy and radar, few physicians had a chance to study microwave bioeffects before the end of the war. During the early years of the war, all information on microwave technology was classified. When restrictions were lifted, there were no spare radar units for use in civilian research. One group at the Mayo Clinic in Rochester, Minnesota, had eagerly followed the development of radiowave technology in the late 1930s and been promised access to the new short-wave generators as soon as their power outputs had been increased. When the war started, the Mayo group discovered that the new equipment became unavailable. Only later did they learn of the secret research projects that kept all microwave generators off the market. They did not succeed in getting a microwave generator from the Raytheon Company until June 1946.

Subsequent studies carried out at the Mayo Clinic left no doubt that the new microwave generators provided an effective tool for inducing local heating. Microwaves could be more easily focused than the radiowaves used in standard diathermy units. They were also reported to be more readily absorbed by the body, thus increasing their efficiency, and they could be easily directed to specific parts of the body. In brief microwaves seemed to provide an excellent means for pinpointing heating effects 'in the manner of a spotlight, thus facilitating clinical application." "Heating by microwaves," the Mayo Clinic researchers confidently announced, offered "the promise of considerable usefulness in the practice of physical medicine."[10]

Before this promise could become a reality, more information on bioeffects and potential hazards had to be gathered. Researchers knew that the body can handle some excess heat by natural mechanisms that increase blood flow and produce sweating. Blood flow is the principal mechanism used to transfer heat from the central core of the body to its surface. Sweating effects surface cooling through evaporation. Thus, when an excessive amount of internal heat is generated by vigorous exercise, the body gets rid of the excess heat by moving it to the surface with increased blood flow and then getting it away from the surface through sweating and evaporation.

This cooling mechanism is not equally developed in all parts of the body, however. Some parts have ample blood flow and therefore the capacity to handle excess heat easily. Other parts, called avascular parts, have relatively limited blood flow and therefore less capacity to deal with excess heat. From this fact it reasonably followed that if microwaves were going to affect the body adversely, they would most likely do so where there is the least blood flow, assuming that heating is the primary bioeffect. This line of reasoning drew the attention of a number of researchers to two critical parts of the body that have relatively inefficient cooling systems, the eyes and testes.

Other factors besides potential overheating made the eyes a logical target of bioeffect studies. Researchers had demonstrated by the late 1940s that infrared, ultraviolet, and ionizing electromagnetic radiation can produce cataracts. If these results were generalizable to all electromagnetic radiation, then it followed that microwaves might also produce cataracts. In addition, since one of the most useful applications of microwave diathermy appeared to be in the treatment of sinus problems, concern arose over the effect that irradiating the sinuses might have on the eyes.

Initial research on potential eye effects done immediately

after the war appeared to set aside these concerns. Studies done at Northwestern University Medical School in 1947 succeeded in focusing microwaves directly on the eyes of dogs and found no adverse effects. Local heating did occur but without producing any noticeable immediate effects. If these results held true for humans, then it appeared "that this method should be a safe and excellent means for the application of localized heat to the eye."[11]

Other researchers were not so sure, however. One group at the State University of Iowa that was looking into the biological effects for a nearby radar subcontractor, Collins Radio, felt that microwaves might have a delayed effect on the eye. To test this hypothesis, they subjected rabbits either to one brief but relatively high-power exposure or to several lower-power exposures and then waited to see whether problems developed over time. Collecting data over time did produce significant results. The rabbits that were given one intense exposure began to develop cataracts three days later. The rabbits given several low-dose exposures developed cataracts as long as forty-two days later.

Even these results did not put a complete damper on the search for medical applications. The Iowa team believed that "since other workers have shown that the microwave generator serves an adequate purpose in medical therapy by inducing temperature increases in selected areas of the body, the findings of this report should not in any way discourage its employment in those areas as a thermogenic device." But hazards had now been shown to be a definite and physiologically precise possibility, which led to the conclusion "that precautionary measures may be of value to workers and patients frequently exposed to the radiations of microwave generators."[12]

When the Iowa researchers turned their attention to the testes, they again discovered adverse effects. This time rats were used to test for bioeffects. The results demonstrated that at temperatures above 35°C, all irradiated rats suffered tissue damage in the testes. At moderate temperatures (31° to 35°C), about one in two animals showed evidence of tissue damage. Below about 30°C, no effects were observed. Once again these results provided specific evidence of possible adverse effects and called for caution; "precautions should be taken by those working in the field of high frequency electromagnetic generators and by those giving treatments with microwave generators."[13]

The Iowa studies and similar ones at other medical centers did not require any immediate shift in attitude toward microwaves. Most of the studies conducted in the late 1940s and early 1950s used high doses of microwave energy to produce unmistakable thermal effects. Admittedly the thermal effects now being reported were more specifically defined than had been the case in the 1930s, when headaches and other fever-related symptoms seemed to be the worst that could be expected. But the more recent studies did not duplicate actual human exposure conditions. They were not done on humans and did not use power levels that commonly existed in the field or in the doctor's office. The majority of early postwar evidence on hazards showed what might happen if exposure levels became excessive, not what was happening under present exposure conditions.

Two factors changed this situation. During the war the microwave equipment in use had power outputs (measured in watts) that were roughly equivalent to the power outputs of diathermy equipment, typically measuring in the tens to hundreds of watts. By the early 1950s the radar equipment being planned and put into operation was delivering peak power outputs that measured in the millions of watts, with corresponding average powers ranging from tens to hundreds of thousands of watts. Within ten years the power of field radar systems increased a thousand-fold or more. This increase clearly invalidated comparisons to medical diathermy and called for new reference points for dealing with the potential hazards of radar systems.

The second factor that turned the hazards problem from a future- to a present-oriented one was evidence of possible radar-related injuries assembled by John McLaughlin, a doctor at Hughes Aircraft Corporation in Culver City, California. Late in 1952 McLaughlin spent time investigating a case of internal bleeding diagnosed in a worker at Hughes. When a hospital spokesman told McLaughlin that the patient's symptoms resembled a picture of mild radiation sickness and asked him to look into it, he did that.

The results of McLaughlin's investigations raised concern. He estimated that between late 1952 and April 1953 he discovered 75 to 100 additional cases of bleeding. He also saw cases of possible cataract formation in the Hughes population. Contacts with colleagues turned up two reports of leukemia in a pool of 600 radar operators, two cases of brain tumors in a five-man team of microwave researchers, a high percentage of jaundice in exposed air force personnel, one additional case of cataract formation in a radar operator, and general complaints of headaches by those working in the vicinity of microwave radiation.

Subsequent searches through the scientific literature led McLaughlin to a small group of articles that made links between microwave exposure and possible health effects and suggested different mechanisms by which microwaves could affect the human body. This information, when combined with clinical evidence, raised the possibility of links between exposure to microwave radiation and health effects. When he brought this evidence to the attention of his superiors, McLaughlin was told to put it all down in writing, which he did in a report made public in February 1953.[14]

McLaughlin's findings produced a genuine dilemma for those who had to make decisions about radar and possible hazards. Few suspected that large-scale problems did or could exist, but the evidence in the report raised doubts. Officials at Hughes were "concerned with Dr. McLaughlin's findings. We hope that he is wrong. We don't know. We are in a position of neither supporting him, nor saying that he is wrong. We just plain don't know. . . . We are looking to somebody to tell us that we have nothing to be concerned about, or, if we have, then what we should do to protect our employees."[15]

The obvious place to turn for advice was the military, since most of the work at Hughes and at other radar factories was being done under military contracts. When copies of the McLaughlin report made their way to those agencies contemplating the new, high-power radar systems of the future, it produced an immediate response. The bioeffects problem became "a fairly hot potato" that could not be ignored.[16] Within months two major conferences were convened, and a full-scale effort to study RF bioeffects was taking shape.

RF Bioeffects as a Military Problem

The military role in the microwave debate has been harshly criticized. Critics have seen in the fusing of planner, producer, user, and monitor in one agency, the military, the potential for serious conflicts of interest. There were, however, no practical alternatives to the course of action that was pursued. No agency other than the military had the resources, interest, and authority to explore the potential dangers of equipment that was so clearly and exclusively military equipment. Few persons outside the military even imagined that a potential problem might exist. When the military took up the microwave debate in the early 1950s, the bioeffects problem was a military problem.

In addition those who turned to the military in early 1953 for a resolution of the dilemma posed by McLaughlin's report did so in part to avoid conflicts of interest. Many persons who were concerned about possible radar hazards felt that the medical community might be reluctant to take evidence of hazards seriously, given their firm belief in the therapeutic benefits of diathermy, a suspicion not totally unfounded. A letter to the *Journal of the American Medical Association* in October 1952 asking whether the cataracts discovered in a twenty-two-year-old radar repairman could have been caused by microwave exposure brought forth a rigorous denial of any connection: "Answer—Radar waves are completely absorbed by the cornea and have not been reported to be a cause of cataracts."[17] This reply was written four years after both the Iowa and Mayo Clinic groups had issued their reports on microwave cataracts in animals and followed the publication of a case study of cataracts in a thirty-two-year-old radar serviceman.[18] If this reply were indicative of the attitude of the medical community, clearly their advice on hazards had to be taken with some caution.

Moreover the industrial representatives who raised the hazards problem in 1953 had doubts about their own ability to solve this problem. Hughes officials felt that they were not "in any position to undertake a [hazards] study" although they were "glad to make available to anyone who does undertake the study whatever facilities" they had.[19] Industry was also subject to the same conflicts of interest plaguing the medical community. In an in-house memo attached to his 1953 study, McLaughlin told Hughes officials that the Raytheon Company, a major manufacturer of diathermy equipment, was upset by the adverse publicity caused by the publication of reports of microwave cataracts and put pressure on the navy to discontinue funding the research that had led to the reports.[20]

The Raytheon Company denied exerting such pressure, and there may be nothing to the claim.[21] But even if the claim is without foundation, it is true that the potential conflicts of interest apparent in 1953 were ones that involved a split between an application-minded medical community and hazards-conscious radar developers, not between a radar-minded military and an environmentally conscious public. Within this setting the shift from medical to military spheres that took place in the decade surrounding 1950 was neither irrational nor irresponsible. If radar and the new microwave technology posed a threat to exposed populations, the research branches of the military were the most likely bureaucratic organizations to turn to for an evaluation of potential hazards.

Radar was used by all three branches of the military, so any one could have been the reasonable choice for lead agency. The two most likely candidates, since they had the best-established research programs, were the navy and air force, and it was at their initiative that conferences on microwave bioeffects were convened shortly after McLaughlin's report made its rounds in Washington. Both branches remained heavily involved in the RF bioeffects field thereafter, accounting on an average for about half or more of the support that has been available annually for research.

Assigning the problem primarily to these branches did not iron out all of the organizational problems because each wanted to control the program. Between 1953 and 1956 internal bargaining went on at the highest levels of the military to determine which service would control the field and whose labs would receive the most support. By 1956 the air force had essentially won the battle. According to one official version, at this time the navy voluntarily gave over its control to the air force and closed out its in-house research effort on microwave effects at the Naval Medical Research Institute.[22] According to unofficial accounts, the transfer was accomplished only after a great deal of arm-twisting and was not greeted with much enthusiasm by the navy.

A second organizational problem that plagued the military in the mid-1950s was the potential for disagreement between its operations and research units. Operations personnel were concerned with getting microwave equipment into the field and improving defense capabilities. The research branches tended to be more interested in health problems and basic research questions. These differing interests could cause problems. In the late 1950s when researchers finally came up with a guideline for regulating exposure, the operations branches were reluctant to accept it; "could you have heard the protests of our operational colleagues when they first were told to live with this level, I am sure you would have concluded that operational suitability was not the basis for" the proposed guideline.[23] These and other internal conflicts have kept the military from acting with one mind on the bioeffects issue.

Eventually the organizational problems were overcome. In late 1956 military planners, working with advisers from George Washington University, pieced together an agreement that allowed the three services to continue bioeffects research, now under air force direction, and located the majority of the research effort at a major operations center, the Rome Air Development Center in Rome, New York. With all of the services participating, none was technically excluded from the effort, although the air force, under Colonel George Knauf, clearly exercised the leadership role it had been assigned. Locating the research effort at an operations center kept hazards investigators in close contact with radar developers. By early 1957, funds had been committed, in-house research programs initiated, and contracts signed with university investigators. The microwave debate had become officially a military problem.

The centralization of most bioeffects research in the Tri-Service Program crystallized changes that had been ongoing since the late 1940s. Military planners were concerned primarily with the hazards issue, not medical application. Research on therapeutic application therefore gradually disappeared and was replaced by hazards studies. As medical application was slowly pushed into the background, fewer and fewer clinical physicians spent time investigating RF bioeffects, their places taken by research biologists and engineers who tended to pose the hazards question in more purely scientific terms. Eventually the attention of the professional bioeffects community came to focus on one overriding research issue, that of basic mechanisms. If through pure scientific research scientists could discover how RF energy interacts with living tissue, then firm, scientific statements could be made about the possibility of hazards and the RF bioeffects problem solved.

Throughout this transition one key factor remained unchanged; the majority of doctors who came to bioeffects through an interest in diathermy firmly believed that the only effects that needed to be taken seriously were thermal effects.

Similarly most of the biologists and engineers who worked to solve the hazards problem during the Tri-Service era (1956–1960) were preoccupied with thermal effects. Between 1930 and 1960, the personnel, funding sources, and research protocols of the RF bioeffects field changed dramatically, but the scientific mentality that underlay these changes did not.

Tri-Service Research Program

Radiowaves and microwaves come in many different forms. Their wavelengths, which are inversely proportional to their frequencies, can vary from relatively long radiowaves to shorter microwaves. The shape of each wave can be smooth or rippled (amplitude modulated). The distribution of waves over time can be even (continuous wave or CW) or come in spurts (pulsed). And each wave, whatever its form, can be beamed at different power levels. Each variation could have a different effect on living tissue, and each could require a specific experiment to detect its effects, thereby potentially calling for an endless series of experiments to determine the impact of RF radiation on human health.

The organizers of the Tri-Service program decided to concentrate their investigations on the wavelengths commonly used in RF technology: 200 MHz, 3000 MHz, 10,000 MHz, 24,500 MHz, and 35,000 MHz. Their research plan called for exposing test animals to each of these wavelengths in an effort to find out whether one or more produced adverse effects. Rats, rabbits, dogs, and monkeys comprised the mainstays of most experiments. Dogs and monkeys approximated humans in bulk and organ structure, rabbits were used primarily for eye experiments, and rats were chosen for speed and economy in experimentation. The primary direction given to the researchers who received Tri-Service contracts was to find out what happened to these animals under various exposure situations.

Within this general experimental framework, more specific decisions had to be made. Researchers knew that at high enough power levels, they could "cook" experimental animals in a matter of minutes or even seconds. They also suspected, although this had not been proved at the time and is still disputed, that animals could be exposed at very low power levels ad infinitum without inducing any adverse effects. Between these two extremes of known danger and presumed safety there existed an area of potential or possible risk. Before researchers could conduct research in this area, they had to set its boundaries. In other words, before actually undertaking RF bioeffects research, decisions had to be made on where to look for bioeffects.

Tri-Service planners relied on thermal thinking to set exposure boundaries for subsequent experimentation. Working back from the assumption that the only effect that RF radiation has on living tissue is heating, they attempted to estimate the point at which RF radiation adds more heat to the body than the body can normally dissipate. They used mathematical calculation, studied the results of past experiments, and conducted thermal stress studies.

The results of the preliminary estimation efforts consistently pointed to exposure levels between $10mW/cm^2$ and $100mW/cm^2$ as the critical area that needed to be investigated. Exposure levels above $100mW/cm^2$ were known to supply more energy to the body than could be dissipated. The resulting heat buildup obviously could be hazardous if it became excessive. The energy delivered to the body in the $10–100mW/cm^2$ range seemed to fall into a tolerable range. It did not appear to overload the body's normal cooling system. On the basis of this information, the major focus of the Tri-Service program became spelling out in as precise scientific terms as possible exactly what happened to the body when RF radiation initiated thermal stress at about the $100mW/cm^2$ level. This task was assigned to investigators at the University of Buffalo, Massachusetts Institute of Technology, University of Iowa, Rochester University, Miami University, and other schools, each of which received funding from the military during the Tri-Service era.

The thermal orientation of the Tri-Service research is evident in the work carried out under the direction of one typical and influential investigator, Sol Michaelson, at Rochester University. Prior to the Tri-Service era, Michaelson, who was trained as a veterinarian, had studied the impact of ionizing radiation on animals. When the need to study nonionizing radiation bioeffects arose, he was in an ideal position to extend his work to this new field of inquiry. As a result he became one of the first recipients of Tri-Service funding in late 1956 and early 1957. His switch to nonionizing research was effected by gaining access to one RF generator at the University of Buffalo and being granted permission to use an AN/MPS-14 radar set housed at the nearby Verona Test Site at Griffith Air Force Base. The Buffalo generator produced 200 MHz CW radiation while the Verona set generated 2800 MHz pulsed radiation, giving Michaelson's group the ability to test the impact of several different combinations of RF radiation on his experimental animals. The latter included rats, rabbits, and dogs.[26]

Michaelson began his RF studies by testing how different animals responded to known thermal doses of RF radiation. Dogs, cats, and rabbits were placed in exposure chambers and subjected to incident power levels of 165mW/cm^2 for a specified length of time. Rectal temperature readings were used to monitor how animals handled the excess heat load produced by the RF radiation. The results showed that rabbits, with their thick fur and sluggish circulatory system, experienced a rapid rise in body temperature, developed spasms, and died within about one-half hour after the onset of exposure. Rats experienced the same effects, although over a slightly longer time frame, while dogs seemed to be able to adapt to the excess heat. The rectal temperatures of dogs rose rapidly at first, then equilibrated for a period of time at less than lethal levels. Only after prolonged exposure (one to five hours, depending on the frequency used and exposure conditions) did the equilibrium (phase II) break down and temperatures resume their rise to the critical point (figure 2).

Having established the general characteristics of RF-induced thermal effects, Michaelson designed other experiments to determine how the excess heat affected the body. He first compared RF heating to environmental heat stress by placing other animals in high-temperature, high-humidity (120°F and 50 percent humidity) environments. Similar temperature rises were recorded, with the only difference being that there was no phase II equilibrium period, even in dogs. The body temperature of dogs in the hot, humid environment rose steadily, whereas in an RF field it rose, reached a plateau, and only after a period of time continued on an upward course. Apart from this difference the response to environment alone seemed to be similar to the response to RF radiation.

Once the general thermal response of animals to high-level RF exposure had been established, Michaelson turned to other experiments. Animals were anesthetized with phenobarbital and then exposed. Blood tests were run. Specific parts of the animals' bodies, such as the head, were irradiated. The effect of

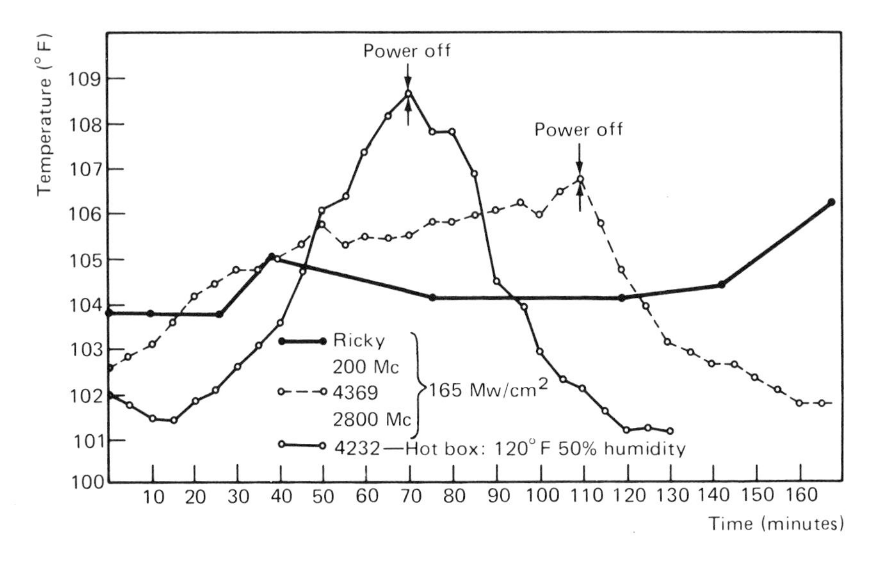

Figure 2
Comparison of thermal responses of normal dogs. From a report by the University of Rochester group (S. Michaelson and colleagues) in the *Proceedings of the Third Annual Tri-Service Conference on Biological Effects of Microwave Radiating Equipment* (1959), p. 175.

RF radiation was tested in combination with ionizing radiation and drugs. Lower exposure levels were used (100mW/cm^2) and the results compared to higher ones.

The understandings that emerged from Michaelson's thermal studies tended to support the common belief that overheating from RF exposure was a problem but not insurmountable or catastrophic. When exposure levels were lowered from 165mW/cm^2 to 100mW/cm^2, the body temperature of exposed animals increased less rapidly. The lesser increases in temperature could be tolerated for much longer periods of time. At 100mW/cm^2 the body temperature of dogs rose and then stabilized for the duration of the test (six hours); thermal breakdown did not occur, and after exposure, all vital signs eventually returned to normal. These and similar experiments seemed to demonstrate that RF bioeffects were a short-term, reversible phenomenon that, if kept in check, presented no problems. As long as critical thermal thresholds were not reached, RF radiation seemed to pose no unusual health-related problems. Once again a thermal solution to the RF bioeffects problem had been found.

There was more to Michaelson's work than this simple solution suggests. His experiments also produced some surprises. When the exposure conditions in the dog experiments were altered so that only their heads were irradiated, the Rochester group found, somewhat unexpectedly, that whole-body temperatures still rose very rapidly and were not accompanied by any phase II equilibrium. Although the energy levels used were large, it was difficult to explain these results on purely thermal grounds, especially given the blood flow patterns. Could the energy received in the head region be transferred to give a whole-body temperature rise? Or did this experiment perhaps indicate, as Michaelson himself suggested, that "at certain levels of exposure to microwaves there may be direct brain stem effect[s]"?[27]

Whether there were such direct brain stem effects, researchers did not know. They also did not know how important a factor time was. Michaelson's group presented some evidence that high-level, short-term effects could be duplicated by lower-level, longer-term effects, suggesting that the critical factor in considering effects (E) might not be incident energy (I) alone, but incident energy times exposure duration (T), or $E = I \times T$.

How far this line of reasoning could be carried raised additional questions. Most researchers felt that there was a threshold power level I_x below which hazards or effects would not occur, no matter how long the exposure period: $E \neq I_x \times T$. No evidence was advanced to support this assumption, and evidence did exist that raised doubts. Thus the thermal solution was not above question.

Few questions, however, were raised at the time regarding the validity of the thermal solution. Researchers had not demonstrated that RF energy did anything but heat the body. And if overheating were all that mattered, then at some point the body's natural cooling mechanisms would protect it from the potential dangers of RF exposure. Once this point of natural defense was found, the military lost interest in the bioeffects problem. It terminated the Tri-Service program in 1961. As far as the official record was concerned, no one had died from exposure to RF radiation, and there was no compelling evidence to suggest that any serious adverse effects were being experienced. That this should have been the case was not surprising if the problem being confronted was one of simple overheating.

As logical as the thermal solution to the RF bioeffects problem seemed in the late 1950s, or earlier in the 1930s, it was not based on the sort of critical reasoning that is supposed to typify the scientific method. It was commonly argued at the time that because thermal effects disappeared below some threshold, all effects must disappear below that threshold. This conclusion would follow if it were known that there could be no effects other than thermal effects, but this was not known. Very few experiments had ever been done below the presumed thermal threshold to see whether any effects appeared. In the early years of RF bioeffects research, scientists did not test the thermal solution; they sought simply to clarify it. In so doing they limited their scientific activity to the least imaginative and most conservative lines of investigation.

Critics of past microwave policy have seized on methodology as evidence of a conspiracy to withhold evidence of low-level hazards from the public. Although there have been conspiracies to withhold information from the public, it is not so much the conscious as the unconscious impediments to breaking out of the thermal mentality that have exerted the greatest influence on the RF bioeffects field. Ideally scientists are supposed to be intellectually curious creatures who leave no stone unturned in their search for truth. In practice, however, they are commonly normal human beings who too frequently follow the path of least resistance. It took extraordinary pressure from outside the domain of science to break the stranglehold the thermal solution had on the RF bioeffects field and open scientific thinking to another class of phenomena, known as athermal or nonthermal effects.

3

The Search for Standards

It would be highly desirable in the light of these observations to set about establishing standards for the protection of personnel exposed to intense microwave radiation before anyone is injured. We have here a most unusual opportunity to lock the barn door before, rather than after, the horse is stolen—J. W. Clark, RAND Project Report, U.S. Air Force, March 7, 1950

The most important problem that has preoccupied the RF bioeffects field over the past four decades has been the search for standards. If scientists had discovered how much RF radiation is hazardous to humans when they learned to exploit radiowaves and microwaves for technological purposes, there would be no microwave debate. Setting RF standards has proven to be a herculean task, however. Policymakers have not been able to reach any definitive conclusions about hazardous or safe exposure levels. The federal government has no general population RF exposure standard, its occupational exposure standard was withdrawn, and only a fraction of the electronic devices that emit RF energy are regulated by federal law. The widely used voluntary standards have no legal power, have had their objectivity questioned, and may be applicable only to select populations. Four decades and millions of dollars of research have not produced satisfactory RF standards.

The usual reason advanced to account for the lack of RF standards is the incompleteness of the scientific record. The logic behind this argument assumes that if enough scientific evidence on RF bioeffects were collected, a comprehensive standard could be set. In the absence of enough evidence, the standard itself becomes elusive. As appealing as this explanation is, it captures only one of many complex factors that have made the search for an RF standard difficult. To be sure, more and better scientific evidence would help answer some questions, but more evidence alone would not resolve all of the problems that have plagued standard setters. The problems raised by the bioeffects issue encompass much more than straightforward scientific understandings.

Early Thoughts on Standards

When standards first emerged as a problem in the late 1940s and early 1950s, attitudes toward them were heavily science oriented. The accepted way to determine how much exposure was dangerous was to begin by collecting scientific data. The scientists who collected the data were routinely asked to interpret them and set standards. Standards in turn were seen as objective reflections of scientific knowledge. When air force officials began to take the RF bioeffects problem seriously in early 1953, they turned to scientists at their Cambridge Research Center for answers. The center's mission was expanded "to include research and development in the biological aspects of microwave energy." The objective of the expanded research was to examine existing information and assess the extent of the problem, establish "research projects to solve the problems identified," and "determine permissible dosages of microwave radiation to include single as well as repeated exposures." Through this simple procedure—identify the problems, undertake scientific research to solve them, set standards—the standards problem was to be solved.[1]

This straightforward scientific approach to standards lay at the heart of all decision making during the Tri-Service era. Scientists were handed the task of identifying and explaining bioeffects. When the scientific data they collected seemed adequate for setting standards, standards were set. By 1960 the three branches of the military and their industrial contractors had concluded from the scientific investigations undertaken that 10mW/cm^2 was a safe exposure level, set this standard, and handed the problem of public sector protection over to a private standard-setting organization.[2]

Even as the 10mW/cm^2 standard was being formulated and established, there were those who recognized that standard setting was not as simple as it appeared. Doubts about a straightforward scientific approach were raised as early as April 1953 by participants at a conference in Bethesda, Maryland, called by the navy to discuss the potential hazards discovered by McLaughlin. During the Bethesda conference, several important and perplexing issues were touched on, all of which have subsequently proved troublesome for standard setting.

Participants at the 1953 navy conference were well aware that science does not always yield unambiguous information. Evidence on health hazards is frequently derived from animal experiments. Extrapolation from these experiments to humans brings uncertainty into the scientific process: "Studies of animals are single studies, and the principle of extrapolation is difficult from animal to the human body."[3] Direct human data can also be problematic. Human data collected under field conditions—such as persons working in a radar field—are difficult to control and usually not replicable.[4]

Scientific experiments require interpretation, but two researchers can interpret the same data differently. During World War II one researcher who surveyed the health of a few dozen radar workers saw in the figures he reported no unusual health effects. Ten years later McLaughlin reexamined the same data and reported to his colleagues at the navy conference that he could not agree that the findings were normal.[5] Results that are interpreted by one researcher as evidence of effects can be interpreted by another to be evidence of no effects, and these interpretations will influence the standard-setting process.

The problem of subjective interpretation led one participant at the conference to suggest that it might be necessary to assemble a neutral board to "review the literature and . . . form some sort of opinion from what already has been done." The call for a neutral board was prompted by a desire "to get away from the colored interpretation that the investigator himself makes for his paper." As reasonable as this suggestion was, it raised another problem: "The only people in the field thoroughly familiar with it are the people who have raised the doubts about it."[6] There were no outside observers to staff a neutral board for the purpose of securing an objective review. Thus, it is not only the subjective or biased interpretation that is a problem, but the fact that bias is not easily overcome since the reviewer and the person or work being reviewed often represent the same school of thought.

Policymakers almost always take more than science into consideration when they set standards. Early efforts to set RF standards had to take into account the attitudes of persons in operations, some of whom vigorously opposed the idea that radar could pose a hazard: "there are some who pooh-pooh it, and when we first conjectured that it [i.e., radar] might constitute a hazard, we met opposing views."[7] The persons charged with setting standards knew this, knew that there would be pressure exerted to institute the least restrictive standard possible, and knew that they would have to have solid evidence and a great deal of bureaucratic clout if they were to succeed in instituting any regulations.

Operational pressure was not the only nonscientific factor that influenced standard setting. Participants at the navy conference understood that once a standard was set, "probably a lot of people can prove that they have received over this limit in the past, and could set that up into a law suit for damages."[8] The simplest way to avoid lawsuits would be to resist setting standards below existing exposure levels, but resisting lowering standards for legal reasons would again bring nonscientific pressures to bear on standard setting.

Once the scientific ambiguities, interpretive difficulties, and external pressures have been duly noted, there remains the problem of basic philosophies. In setting standards policymakers have to decide who is to be protected from what and to what extent. This problem too was recognized and discussed in 1953. McLaughlin aggressively set out the most conservative philosophy possible, that of zero risk: "Why put any standard? Why not say the people will not get in the radar beam?" The practicality of his view was immediately challenged:

Commander Brody: We may have a mechanic working on the landing gear. He doesn't have anything to do with the radar.

McLaughlin: Then the radar shouldn't be turned on.

Brody: These are big aircraft, and twenty people buzzing around it [*sic*]. Surely it is quite conceivable [for there to be exposure] and much of this would be inadvertent exposure.

McLaughlin: Point it where no one is in the range. Why mention any level at all? Just say, "You will so arrange your work so that no one is exposed to the radar beam, period!"

Brody: People are always getting hurt in the service.[9]

Some people are philosophically disposed to accept risk; others would rather not.

If the zero-risk philosophy is rejected, which it almost always is, then decisions have to be made about how much risk to accept. Even here attitudes can vary. Most commonly policymakers work on the principle that if errors are made, they should be made on the side of safety. This is why one participant at the navy conference strongly opposed the incorporation of a time factor into any standards that might be set: "So I say, be overly cautious, recognize the fact [that radar does pose a hazard] and come out flatly and [set a] reasonable power output without specifying time. . . . In that way, we can err on the side of being ultra-conservative, and hope we won't get into any trouble as to the biological effects from its use."[10] But not everyone agreed with this point of view. Preferences vary depending on how much risk is deemed acceptable, and this in turn affects the level at which standards are set.

This list of factors affecting standard setting does not exhaust the richness of the discussion at the 1953 Bethesda meeting. The participants knew that funds, peer pressure, and the implications of experimental results would have bearing on the course that science took. Although none of the participants was a trained policy expert, they also understood that they were dealing with a potentially explosive public issue, one that they did not want to "appear in the Washington Post . . . tomorrow."[11] Yet they were confident that an honest, joint, interdisciplinary effort could achieve an RF standard.

Philosophy of the Military's 10mW/cm² Standard

Whether subsequent policymakers undertook a proper course of action in setting standards can be questioned if their actions are judged in accordance with the understandings evidenced at the 1953 navy conference. When theory turned to practice and recommendations became decisions, most of the factors discussed at the conference were quickly forgotten. Conflicting points of view were passed over, scientific ambiguity was ignored, and contrasting philosophies left unexplored as a single-minded approach gradually crept in and came to dominate all decisions.

In the press to get interim guidelines in place and to establish permanent standards, the military forgot about neutral review boards and objective interpretations. Instead of broadening the base of decision making and encouraging conflicting points of view, military planners gradually vested most authority in the hands of a few like-minded individuals. During the early years of the Tri-Service effort, George Washington University consultants were hired to provide outside opinions. Later this source of diversity was dropped as the responsibility for the Tri-Service effort came to rest first in the hands of the Ad Hoc Tri-Service Committee on Biological Effects of Microwave Energy (1958) and eventually of one man, Colonel George Knauf.

Knauf was a surgeon familiar with military-related medical problems. After service during World War II and in Korea, his attention turned to the bioeffects problem when he was assigned to the Electronics Research Center at Rome Air Development Center. Here he came in contact with the latest high-powered radar systems and the problem of potential hazards. Once convinced that there were issues that needed to be addressed, Knauf set about organizing the Tri-Service research program at Rome.

Knauf's influence on the RF bioeffects field extended beyond organization of the program. During the years of the program, he not only kept close track of the research being conducted but made many of the major decisions on interpretation and application. In essence Knauf became the conduit through which all scientific information had to pass as it made its way into the realm of policy and standard setting. As this occurred, the diversity built into the research program was lost. This was especially true in the discussion of RF exposure standards.

The discussion of RF exposure standards began in earnest in April 1953 with the convening of the navy conference at Bethesda. The participants at this conference did some rough calculations on thermal overloads and concluded that 100mW/cm^2 was a probable upper limit for safe exposure. But some of the assumptions made during the conference were erroneous, a fact that prompted one participant, Herman Schwan, to do some recalculating after the conference based on more accurate information. Schwan's recalculations placed the level for probable safe exposure—the point at which the heat added to the body could be handled by the body's normal cooling system—a factor of ten below the earlier figure, or at 10mW/cm^2. He forwarded his recalculations to the navy in May 1953, one month after the conference.[12] Thereafter 10mW/cm^2 became the focus of almost all subsequent discussions of standards.

The air force, which set the pattern for most military policy in the 1950s, made several key decisions on the acceptability of the 10mW/cm^2 figure as early as 1954, well in advance of the start of its own research program and the Tri-Service effort. By May–June 1957, still one month before the first Tri-Service conference (July 1957), the air force had adopted 10mW/cm^2 as its working standard for all field units.[13] In other words the basic pattern of air force thinking was developed in advance of the basic research that was designed to set standards and without taking into consideration the potential advice of the Tri-Service program.

Once the Tri-Service program moved into high gear in 1957, Knauf's role as interpreter and judge became apparent. As he surveyed the preliminary reports presented during the annual Tri-Service conferences, he became convinced that none of the work undertaken invalidated the 10mW/cm^2 standard. Two years into the program he reassured participants at the second Tri-Service conference, "I think this might be a good time to say that up to date there has not been any effect produced or even hinted at at power levels which remotely approach our established maximum safe exposure level." Two more years of research did not shake Knauf's conviction. His opening remarks at the fourth Tri-Service conference again reassured participants: "I am indeed pleased to say that up to today we have not seen any research data which shakes our faith in the validity of this arbitrary safe exposure level [10mW/cm^2], which we sponsored some five [*sic*] years ago."[14]

It was such judgments made within limited policy circles that carried the day, not the collective wisdom of the Tri-Service program in general. To be sure, any researcher could have reported or said whatever he or she wanted. Many did express reservations about the adequacy of existing data; a few felt that some scattered anomalous findings deserved more attention. But no mechanism existed for getting these perspectives into the decision-making process. Military officials did not question their own interpretations or worry that they had left no room for viewing the RF bioeffects problem from other perspectives. Rather than guarding against single-mindedness, military planners allowed it to become a fact of life.

The shortcomings of the military's approach to the standards question might be understandable if competing opinions were in relatively short supply at the time, but this was not the case. In the period prior to the Tri-Service program, many different approaches to the hazards question were in the air. Organizations other than the military had to make decisions about exposure levels, and their actions show that the pieces of the standards puzzle could have been put together in different ways.

Industry, for example, initially adopted a more conservative attitude than the military toward the critical issue of margin of safety. The military's 10mW/cm^2 standard was set a factor of ten below the level at which injury might be expected—100mW/cm^2. Officials at the General Electric Company felt that 100 provided a better margin of safety and initially set 1mW/cm^2 as their in-house standard. Bell Labs adopted a safety factor of 1000, placing their initial RF standard at 0.1mW/cm^2.[15]

The reason one of these attitudes came to dominate leads back to the restrictive decision-making procedures adopted by the military and the role played by a few pivotal individuals. GE ultimately came around to the military's point of view as a result of personal contacts between Knauf and GE's main consultant on standards, Benjamin Vosburgh. The critical negotiations between these two men came in 1958 after both had heard rumors about each other's position. Knauf heard "that G.E. was unwilling to accept [the military's] proposed safe exposure level." Vosburgh understood that the military was thinking of abandoning the "10mW/cm^2 level for a more conservative one."[16] Both rumors implied that there was some desire for a lower standard, a fact that concerned Knauf. He had already gone on record as saying that "it is immediately apparent that all three [standards—GE's, Bell Labs's, and the military's] cannot be right."[17] By implication a lowering of the standard would have meant that the military, not industry, had been wrong. Knauf discussed this situation with Vosburgh and shortly after GE relaxed its standard, as did Bell Laboratories. By the time of the second Tri-Service conference, Knauf could confidently announce, "I assure you that not only have we not had any real differences, but in truth have been able to work together in a delightfully harmonious manner."[18]

The compromises reached in agreeing to the 10mW/cm^2 figure had the effect of undermining yet another factor discussed in 1953, the philosophy of optimum safety. Although military and, later, civilian officials have consistently argued that their primary concern has been optimum health protection, their actions point to a different philosophy. At the very least the adoption of a safety factor of 10, instead of 100 or 1000, suggests that policymakers were not being as cautious as they could have been. Most thought that they were being cautious enough; there is little if any direct evidence to suggest that the actions taken knowingly permitted hazardous conditions to prevail. But the fact of the matter is that the early decisions on standards did not provide exposed populations with anything more than the perceived minimum protection necessary to avoid adverse health effects.

Some of the participants in the early negotiations over standards understood that this minimum-protection attitude placed a burden of risk on exposed populations. Vosburgh and GE knew that they were striking an agreement with the armed services; they were not agreeing to a known safe exposure level. Vosburgh even went so far as to suggest that if the standard were set at 10mW/cm^2, "it will become necessary generally to monitor at a 1mW/cm^2 mean watt value in order to make the necessary allowance for harmonies and spurious waves." In other words Vosburgh felt that the 10mW/cm^2 level was close enough to the safety-risk line to require constant monitoring. Moreover he believed that 1mW/cm^2 "provides quite a bit of total body energy exposure when one considers that the average body surface of 2 meters would absorb approximately 10 watts." This amount of input provided no problem from a strict thermal point of view, but the situation might change "if and when it has been proven that some important part of that energy is absorbed by susceptible tissues in the form of nonthermal energy having a cumulative effect" and require reappraisal of the safety factor issue.[19]

One reason Vosburgh and others were willing to relax the margin of safety is that actual exposure levels in their factories were getting near the 1mW/cm^2 level. GE had no trouble keeping exposure below 1mW/cm^2 with radar equipment that operated in the kilowatt range, but as power increased to the megawatt range, this level "became quite difficult to maintain." In some cases whole areas had to be vacated while key pieces of equipment were being tested, raising the possibility of restricting technological development.[20]

The potential clash between technological development and

health protection caused Vosburgh and others to reevaluate their philosophy of risk. How far could one reasonably go in the pursuit of safety? As he justified GE's relaxed attitude, Vosburgh articulated the philosophy that was slowly taking control: "It is reasonable to err on the safe side but not so far that it hurts; not so far that progress in the art becomes jeopardized; not so far that we will one day laugh too loudly at our present day fears."[21] Safety had become a relative factor, an objective to be sought within specified bounds. The philosophy of standard setting had become one of being conservative, erring on the side of safety, as long as it is possible to do so without presenting roadblocks to technological development. The tacit acceptance of this philosophy was the first step taken toward shifting the burden of risk more squarely to the shoulders of exposed populations.

A second step in the same direction is apparent in the way in which Knauf and his Tri-Service program colleagues selectively used scientific evidence for justifying the $10 mW/cm^2$ figure. Tri-Service organizers such as Knauf did not accept all effects as evidence of potential harm; they accepted only immediate permanent damage as significant. Since workers who complained of headaches during exposure generally felt better once exposure stopped, these symptoms were dismissed as of no long-term significance. Minimal overheating was also ignored since the body could cool itself. Testicular damage, although known to occur at or near the $10 mW/cm^2$ level, was not taken seriously. (See figure 3.) One air force paper evaluated the testicular effect as "readily detected histologically, but its importance may be negligible."[22] The only effects given any serious consideration were cataracts, and these seemed to occur only at levels above $100 mW/cm^2$. Having dismissed all seemingly minor effects as trivial, Knauf and others could conclude that they "had not seen any research data which shakes our faith in the validity" of the $10 mW/cm^2$ standard.

The result of this attitude further eroded the margin of safety supposedly built into the $10 mW/cm^2$ standard. It ensured that the burdens of any errors would fall on the exposed populations, not on those who were exploiting microwave technology. When doubts arose, the decisions that followed allowed development to proceed until hazards could be proved. Uncertainty was not taken as sufficient grounds to impede progress. The fact that long-term studies had not been conducted or that

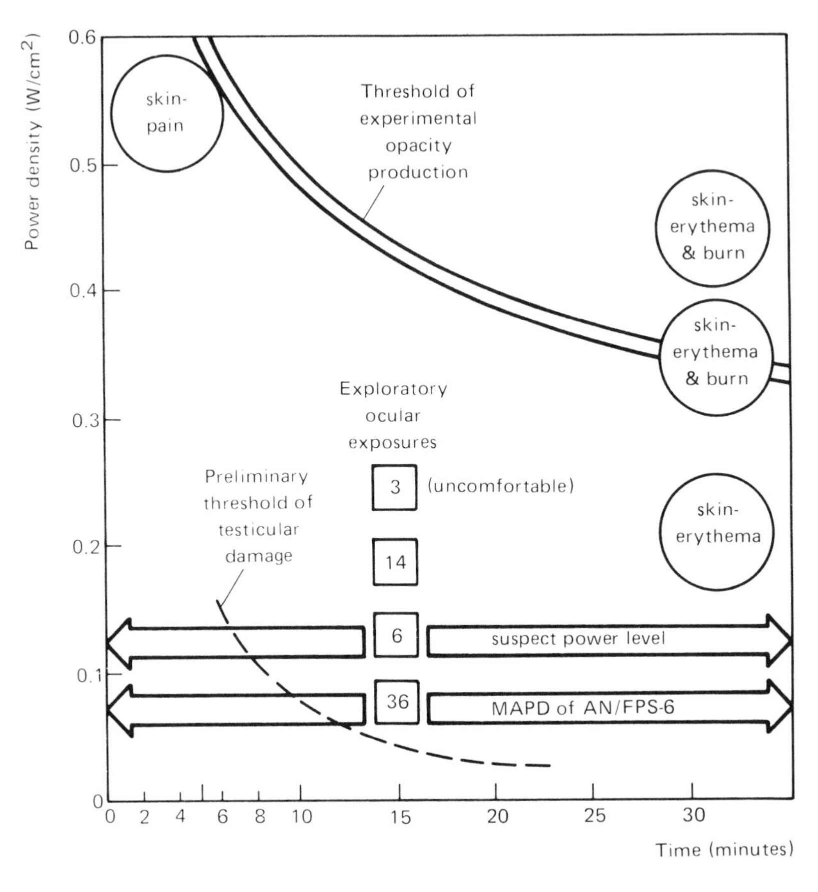

Figure 3
Summary of bioeffects data. From a classified report of the School of Aviation Medicine, U.S. Air Force (September 1955), p. 11.

doubts still existed about the possibility of athermal effects was not deemed sufficient ground to be more cautious by setting lower standards or delaying the deployment of very high-powered radar units. This ultimately was the philosophical approach to standard setting that was assumed over the course of the 1950s.

It should not be inferred that military policy setters knew of undue risks or expected that injuries would turn up. Once they were convinced 10mW/cm^2 was safe, many reasons were found to rationalize the accepted standard. In addition the attention of the age was focused more on development than on health protection. Knauf likened the job of weighing the importance of health protection against the pace of technological development to that of a three-year-old child attempting to follow its father, adding that after the launch of Sputnik in 1957, "the task of keeping up with Dad was just about hopeless."[23] Development was the goal of the age. Environmentalist thinking had yet to mature (Rachael Carson's *Silent Spring* was not published until 1962), and microwaves were still primarily a military technology. Within this historical context 10mW/cm^2 seemed to most to be a reasonable compromise. But the compromise became almost impossible to undo, no matter what evidence or thoughts to the contrary arose.

Philosophy of ANSI's 10mW/cm^2 Standard

By the late 1950s, when the manufacture and use of microwave technology was spilling rapidly out into the industrial and civilian sectors, the need for a broader, nonmilitary RF exposure standard was becoming apparent. If the military had found it necessary to develop standards as a result of a concern over hazards, standards would almost surely be needed for industry as well. The obvious place to turn for uniform, industrial standards was the American National Standards Institute (ANSI).

As important as ANSI has been in the debate over standards, it is important to stress that ANSI does not set standards. Its role in standard setting is that of a facilitator or mediator. ANSI officials will, if requested, determine the need for standards and ask organizations or individuals to develop them. They will also help validate standards through national votes and ensure widespread circulation by publishing the adopted standard as an ANSI standard. But the standards ANSI publishes are simply standards developed by one or more sponsoring organizations and subsequently voluntarily adopted by the community of users.

ANSI was drawn into the microwave debate in May 1959 when the navy's Bureau of Ships asked it to sponsor a general meeting to assess the need for standardization in the field of radio-frequency electromagnetic radiation hazards. After some discussion, the delegates unanimously agreed that ANSI should "establish a project in regard to hazards arising from radio-frequency electromagnetic radiation," that "the sectional subcommittee method" should be used to pursue this project, and that the navy and the Institute of Electrical and Electronics Engineers (IEEE) should cosponsor the project.[24]

As reasonable as these suggestions were, they came at a time when politics and jurisdictional disputes were beginning to enter standard setting. By 1959 factionalism had started to surface within the Tri-Service program as the services vied for control over the bioeffects program and the authority to set standards. Department of Defense orders in 1958 clearly assigned to the air force "standardization responsibility for electromagnetic radiation hazards to personnel." But at precisely the same time the navy's Bureau of Ships was assigned "standardization responsibility for electromagnetic hazards common to personnel, fuel, and equipment."[25] Although the bureau's authority did not encompass the bioeffects program, which remained at Rome under Knauf's control, the navy interpreted its responsibility broadly and began looking at personnel standards. Rather than going to the air force for help, navy officials turned to ANSI.

The working alliance developed between ANSI, at the time headed by U.S. Navy Admiral G. F. Hussey, Jr., and the navy diminished the importance of the air force in standard setting. Colonel Knauf was not pleased with this development and tried to ensure that one branch of the military did not control ANSI's deliberations. However, his lone voice for the air force at a May 1959 ANSI meeting did not outweigh those of the six delegates from the navy in attendance, and shortly after the navy assumed a major role in setting the course of most subsequent ANSI work.

The navy's leadership role not only tested its ties with the air force but also severely strained relations with its cosponsor, the IEEE. In theory the two were supposed to work together in appointing a standards committee and chairman. In fact, even before ANSI had tallied its vote on the question of the need for a standard, the navy's representative, A. L. Van Emden, had proposed three candidates for chairman and recommended early July 1959 for the first meeting.[26] The IEEE was not prepared to move this rapidly. Their representative, J. Paul Jordan, "expressed considerable indignation" when he discovered that he had not been kept informed of the navy's actions. To avoid future lapses in communication, the navy called for a meeting of the two cosponsors in December 1959 at which a procedure was agreed to by which a chairman could be appointed and the first official sectional meeting called.[27]

The December 8 agreement fell apart almost as soon as it was drawn up. The navy understood that it was to contact four persons agreed to as possible candidates. If none of the four was available, then the navy could "select and obtain a suitable chairman."[28] Jordan, however, understood that he would make the appointment and that he had the responsibility for contacting the fourth name on the list, Dr. John B. Russell of GE. When the navy's Van Emden contacted Russell on his own, failed to get his consent, and turned to a fifth, unagreed-to candidate, Herman Schwan, Jordan hit the roof. He had already investigated Schwan's credentials and although finding no fault with his "technical competence" objected to the fact that he was "an individual worker and could not operate well in a group." One doctor who had been at a meeting with Schwan even suggested to Jordan that "very little constructive work could be done when he [Schwan] was in attendance at the meeting since he would accept no compromise to his own ideas. The doctor went so far as to imply that he would be unwilling to serve on any committee chaired by Dr. Schwan." In order to avoid embarrassment to the Department of the Navy and to ANSI, Jordan reluctantly accepted Schwan's appointment, promising to watch the committee's progress very closely. If progress did not follow, Jordan was prepared to request Schwan's resignation.[29]

With this cloud of protest and early maneuvering in the background, ANSI's sectional subcommittee on radio-frequency hazards, designated C95, convened its first meeting on February 15, 1960. Under Schwan's direction C95 drew up an ambitious program. Six subcommittees (C95.I–C95.VI) were set up, with each assigned a specific problem to investigate. Quarterly meetings of the subcommittees were proposed; progress reports were to be issued and deadlines set. Schwan hoped to be able to assemble enough material to begin drawing up standards within a year.[30]

In the years that followed the first C95 meeting, very little went according to plan. The navy and IEEE continued to squabble over the reins of control. In April 1960 ANSI asked the two cosponsors to work out their differences and agree to allow one of the two to take over as principal sponsor. No agreement was ever reached, although the navy eventually became principal cosponsor and over the years has assumed most of the organizational responsibility.[31]

The work of the subcommittees proceeded at a snail's pace;

Schwan's one-year deadline came and went without a second C95 meeting ever being called. Committee and subcommittee members failed to show up for meetings or to prepare reports. When standards were drawn up, months were needed to assemble the votes required for adoption. Overall the work that the military had done in a matter of months and that Schwan thought ANSI could do in a year or two took nearer to a decade to complete. C95's most difficult task, setting a personnel standard, was not completed until November 1966.

C95's work proved so difficult because it was carried out at a time when interest in RF standards was on the decline. Following the adoption of its own standards, the military terminated the majority of its research contracts, bringing work in the bioeffects field to a near standstill.[32] With the end of most bioeffects research (although some secret research did continue), general interest in the bioeffects problem, along with the sense of urgency that had existed a decade earlier, diminished. In the early 1960s few people cared about or saw the need for RF exposure standards, making the work of the subcommittee assigned the problem of human exposure standards, Subcommittee C95.IV: Safety Levels and/or Tolerances with Respect to Personnel, next to impossible.

The early history of C95.IV is a litany of one missed deadline after another. Chairmanship of this key committee was initially taken on by Colonel Knauf, who had been reassigned to Patrick Air Force Base in Florida and was preoccupied with the space program. Knauf managed to sketch out a proposal for a literature search, which, if funded, was to be carried out by a scientist at nearby Miami University, William Deichman. Knauf reported this progress to his subcommittee members at their first meeting on January 18, 1961, and at the same time brought them up to date on the results of the most recent Tri-Service conference (the fifth and last one, held January 16–17, 1961). His report included the now-familiar conclusion that the "research data available to this sub-committee to date provides no reason for the modification of the established $10mW/cm^2$ [standard] as a safe exposure level."[33] In this way Knauf transferred to C95.IV the same focal point for discussion ($10mW/cm^2$) that preoccupied the Tri-Service program and that would come to dominate all early ANSI deliberations.

For the next year C95.IV floundered. In September 1962 Schwan was informed by C95's secretary, Glen Heimer, that he expected the "loss of both subcommittees I and IV by this fall." Heimer did "not know Knauf well enough to understand the reason for his failure to produce."[34] Those who did know him understood that he had lost interest in RF bioeffects. By the time of the third C95 meeting in February 1963, C95.IV had folded and Knauf had terminated all contacts with his subcommittee. The main reason advanced for the lack of progress was the collapse of financial support for bioeffects work with the termination of the Tri-Service program. Deichman's study, which may never have amounted to more than a conversation, was not started, and there were no funds in either the air force or the navy to get things moving again. Schwan therefore dissolved C95.IV and, with the help of an engineer from Bell Labs, William Mumford, undertook to reorganize the effort to set a personnel standard.[35]

The collapse of C95.IV essentially returned the standard-setting process to the single-minded approach of the late 1950s. Schwan, like Knauf, believed in the safety of the 10mW/cm^2 standard; it was his calculations, after all, that had led to the adoption of this figure in the first place. Schwan was also prepared to accept this figure as safe on the basis of existing information. While not denying that more research could be done, he believed that C95 and its subcommittees could not wait for more data before setting standards. Time and time again he urged his colleagues on C95 to make decisions on the basis of what was already known: "The Chairman of C95 pointed out that it is not the function of C95 and its working [sub]committees to undertake research in order to correct deficiencies in knowledge. Its task is to formulate standards based on available information. Hence the working committees' primary task must be the evaluation of pertinent information and the formulation of standards which can be well supported by pertinent literature."[36]

Schwan was willing to make such decisions and he did. Within a few months of the November 1962 collapse of C95.IV, he had assumed the role of acting chairman of this subcommittee, put together a new five-member support team, and drawn up a standard (dated June 24, 1963) proposing, not unexpectedly, that 10mW/cm^2 be adopted to "prevent possible harmful effects on mankind resulting from exposure to [RF] electromagnetic radiation."[37]

The eventual reliance on the thinking of only one or perhaps a few persons meant that the many broad factors that could have had bearing on C95's deliberations, such as the ones raised at the 1953 navy conference, were once again ignored. Schwan was aware of these factors and even listed them for his colleagues at a February 1963 C95 meeting: "1. long-time exposure; 2. short-time exposure; 3. define what is meant by: (a) tolerance, (b) safety factor, and (c) the limit between undesirable effect and no effect; 4. attempt to provide tolerance figures."[38] Few, if any, of these points were actually addressed in the standard first outlined in June 1963 and eventually adopted in November 1966 as ANSI Standard C95.1-1966.[39]

Schwan probably did not believe that in supporting the 10mW/cm^2 figure, he was subjecting exposed populations to undue risks. As he told Michaelson later, "I personally consider the 10mW/cm^2 [figure to be] rather conservative. It provides for an adequate safety factor." Exposure at this level, in his view, would produce "barely noticeable rather than dangerous temperature increase." He thought of the bioeffects issue strictly in thermal terms. He knew of no mechanisms that could account for sensitivities to very low energy flux. He assumed that people were already being bombarded with low-level RF radiation and not suffering ill effects, reinforcing his view that the thermal solution held.[40]

Schwan's thinking was flawed by the same inconsistency as the thermal thinking of the Tri-Service era. The scientists who developed C95.1-1966 had studied and understood thermal mechanisms. The arguments about thermal stress used in C95.1-1966 were based on empirical evidence. But these same scientists could not prove that general population effects were impossible and had not undertaken comprehensive studies to look for them. They also did not know, as Schwan implied, that a significant percentage of the U.S. population was being exposed to low-level RF radiation; figures on actual exposure levels were sketchy and unreliable at the time. But these shortcomings were not taken into consideration. The early standard setters accepted thermal thinking as a fact of science and ignored the weaknesses of their evidence through an act of faith.

This logic easily survived in the policymaking circles of the 1960s because no one stepped forward to challenge it. The majority of the U.S. scientific community was satisfied with the thermal solution. In addition, although consumerism and environmentalism were beginning to develop, it was many years before organizations such as ANSI took notice. When C95.1-1966 was circulated for a vote, the members of C95 were carefully broken down into interest groups to demonstrate the broad base of support the standard supposedly had. C95's voting membership included six producers, seventeen consumers, and nineteen representatives of the general interest. The groups considered to represent consumers were the following:

American Petroleum Institute

Armed Forces Institute of Pathology

General Dynamics

National Aeronautics and Space Administration

U.S. Department of the Air Force, Office of the Surgeon General

U.S. Department of the Air Force, Rome Air

U.S. Department of the Army, Electronics Command

U.S. Department of the Army, Environmental Hygiene Agency

U.S. Department of the Army, Materiel Command

U.S. Department of the Army, Office of the Surgeon General

U.S. Department of the Interior, Bureau of Mines

U.S. Department of the Navy, Bureau of Medicine and Surgery

U.S. Department of the Navy, Bureau of Naval Weapons

U.S. Department of the Navy, Bureau of Ships

U.S. Department of the Navy, Marine Corps

U.S. Department of the Treasury, Coast Guard

U.S. Public Health Service[41]

With the exception of the Public Health Service, none of these agencies could be expected to provide what would soon become known as the consumer point of view. The fact of the matter is that C95.1-1966 was developed primarily by producers for industrial and military users, not by consumers or for consumers.

The major objections that arose during the acceptance of the $10 mW/cm^2$ standard came from persons who worried that it might be too strict. When objections were raised to the use of the term *recommendation* to describe what many hoped would be a binding standard, Schwan explained that considerable negotiation was required to reach a satisfactory standard and that the term *recommendation* had been used to please "several groups, which presently strongly object for practical reasons against any standard but their own." Schwan hoped that by going with the weaker recommendation/standard, he could get it accepted immediately and then later use it to "provide an incentive to eventually replace present practices by those implied by the new standard."[42] This compromise position eventually won, although three members of IEEE voted against the adoption of C95.1-1966, presumably because they still had reservations about the ambiguous language that had been adopted to satisfy those who were unwilling to conform to one uniform standard.[43]

The Army Electronics Command at Fort Monmouth, New Jersey, was also uncomfortable with the restrictiveness of C95.1-1966. In a November memo to the Navy Ship Engineering Center, which had replaced the Bureau of Ships as the principal cosponsor of C95, Fort Monmouth officials told the navy that they would not concur with C95.1-1966; they found it too general and insensitive to design requirements. In particular they wanted to know whether C95.1-1966 was intended to apply to "the design of equipment or the exposure time of personnel operating the equipment."[44] Such objections, which harkened back to the research-operations split in the early 1950s, kept C95.1-1966 from having much impact on the military, which was still refining its own $10mW/cm^2$ standard.

There were also internal problems with C95.1-1966. Under Schwan's leadership C95.IV had produced a standard but many details had been glossed over. One particularly annoying problem, which has continued to plague C95, is that of the target population: who are "personnel"? C95's secretary, Glenn Heimer, understood that C95.1-1966 was designed primarily to be an occupational standard. This meant that C95.IV probably would have to address the problem of a second standard for the general public. Heimer felt that "such a non-occupational exposure standard might be aimed at the industrial or domestic application user and permit a reduced exposure level, perhaps in the neighborhood of $1mW/cm^2$ for continuous exposure."[45] But before this issue could be resolved, Schwan withdrew from active service on C95.IV, leaving the subsequent revision of C95.1-1966 to a new committee and a new era in the microwave debate.

Philosophy of the USSR's RF Standard

Within a few months of the second Tri-Service conference (July 8–10, 1958) and Knauf's reassurance that "to date there has not been any effect produced or even hinted at at power levels which remotely approach our established maximum safe exposure level," the chief state sanitary inspector of the USSR, V. Ahdanov, approved Temporary Safety Regulations for Personnel in the Presence of Microwave Generators.[46] These regulations set tolerance levels well below those established by the U.S. military as safe—1000 times below in the case of continuous exposure for an eight-hour working day.

The existence of two different RF standards raised a rather obvious problem. In declaring that Soviet workers on an average were to be exposed to no more than 0.01mW/cm^2 (also written as 10μW/cm^2) of RF energy, Soviet officials were saying that continuous exposure above this level could be dangerous. (USSR workers were allowed to be exposed to levels as high as 1mW/cm^2 for fifteen to twenty minutes, provided suitable protective measures were taken.) U.S. officials were claiming that continuous average exposure at or below 10mW/cm^2 was safe.

Information on the lower USSR RF standards and similar ones set in other Soviet-bloc countries slowly made its way to the United States during the 1960s. When U.S. policymakers examined the Soviet standards and the scientific record on which they rested, they were appalled. "These standards," Michaelson wrote, "are based on vague 'asthenia' syndromes reported by individuals occupationally exposed to microwave fields. These effects have not been demonstrated by Western investigators."[47] Soviet scientists seemingly were willing to accept subjective workers' complaints of headaches and fatigue as evidence for setting standards; Western scientists were not.

Western scientists also believed that the Soviet animal studies used for setting standards were not rigorous. Western scientists did admit that the Soviets were finding low-level effects: "Some studies indicate that exposure of experimental animals may add to disturbance of conditional responses after several to more than 10 weeks with radiation levels in the order of several hundred microwatts per square centimeter," but this still did not provide the hard scientific evidence of hazards for which U.S. researchers were looking. Since the low-level effects were thought to be reversible, Western scientists saw no reason to equate them with hazards. And if there were no low-level hazards, there was no need for standards as low as 10μW/cm². This logic led to the conclusion that the Soviets were wrong.[48]

Soviet policymakers did not approach standard setting from the Western point of view. Instead of grounding all decisions on hard science, they placed more emphasis on behavioral effects, in part because of the Pavlovian origins of their biology. And perhaps most important, they approached standard setting from the viewpoint of workers rather than industrialists. In combination these factors led to a sharply contrasting philosophy of standard setting and to the adoption of lower standards.

The consequences of the Soviet philosophy of standard setting can be seen most clearly in their attitude toward and interpretation of on-site health surveys. Soviet researchers began looking at the health of radar workers at about the same time as did U.S. researchers, in the early to mid-1940s. Both discovered that radar workers complained of headaches and other asthenic discomforts. U.S. researchers universally dismissed such symptoms as inconsequential and with few exceptions discontinued surveying worker health. In the United States *hazard* came to mean demonstrable and irreversible physiological harm. The Soviets not only paid attention to subjective asthenic symptoms but made them the focus of detailed health surveys. Soviet standard setters regarded discomfort as a significant enough health effect to be classed as a hazard and used as a basis for setting standards.

Once this philosophical approach was accepted, the steps to setting a standard followed a straightforward empirical route. A well-known Czechoslovakian scientist, Karel Marha, outlined these steps for U.S. scientists in 1969 at a symposium in Richmond, Virginia: "At first we elaborated some methods for measuring the intensity of the field. . . . Thereupon we visited factories, broadcasting stations, television and radar centers, and spoke with the people there and performed individual measurements. We visited approximately 200 work places with diverse applications of electromagnetic waves. Furthermore, we also carried out some simple biological experimentation in animals to verify the findings from the plant surveys."[49] The plant surveys turned up the usual symptoms: "pains in the head and eyes and fatigue connected with overall weakness, dizziness, and vertigo when standing for a longer period." These and other symptoms were correlated with the power densities being measured and the appropriate standards set. The goal of the standards was "to prevent not only damage to the organism, but to prevent unpleasant subjective feelings as well."[50]

The acceptance of asthenic symptoms inclined the Soviet-bloc standards toward lower acceptable levels. The Soviets were willing to accept this consequence, at least insofar as their published standards reflected the experimental results being reported in their scientific literature. They did not let the scientific ambiguity of the health surveys or even of some animal studies stand in the way of identifying possible hazards, believing instead, as two Polish scientists suggested in 1976, that "it is far better to present approximative evaluations than to create an impression of accuracy where none can be found."[51] Soviet-bloc scientists were willing to accept more scientific error when setting standards, particularly if accepting error was likely to result in greater protection.

Soviet-bloc scientists have also been willing to accept questionable methodological assumptions if these assumptions result in more protection. When Marha presented his summary of Czech procedures at the Richmond symposium, the question was raised whether Czech scientists actually had shown that workers were getting headaches at $\mu W/cm^2$ levels. Marha responded by noting that the Czech standards were not based on actual exposure conditions but represented extrapolations from effects found at higher levels of about 100–150mW/cm^2. This revelation seemingly brought the U.S. and Soviet-bloc positions into remarkable conformity since U.S. scientists have always maintained that effects can be and are found above 100mW/cm^2. At this level heating can occur, and overheating (fevers) can cause headaches. Both sides seemed to be agreeing on the scientific data on which the standards were based. The differences arose when decisions were made about how to extrapolate from these data to standards.

Soviet-bloc scientists used cumulative models to extrapolate from known 100–150mW/cm^2 short-term effects to long-term standards. They assumed that the most important factor to consider in setting a standard is the total amount of radiation received, not the power of any single dose. If 10 minutes of exposure at 100mW/cm^2 produces a headache, then it must be assumed, using a cumulative model, that 100 minutes at 10mW/cm^2 or 1000 minutes at 1mW/cm^2 might also produce a headache. Using this method and building in safety factors, one rapidly gets down to the μW/cm² range for permissible exposure levels.[52] U.S. scientists, by contrast, have consistently objected to the use of cumulative models in setting standards, primarily because they have yet to be convinced on the basis of replicable scientific data that there are long-term cumulative effects. Instead they have relied on some type of short-term thermal model for setting standards.

To date neither model has been proved to provide an accurate description of nature. Certainly when standards were being drawn up in the late 1950s and early 1960s, either could have been and was supported. The selection of one model, along with a concomitant attitude toward scientific evidence and philosophy of standards, determined which standards were adopted by a given country.

As U.S. scientists and policymakers grappled with these issues, their scientifically oriented attitude pushed standards toward upper limits—to the point nearest to clearly demonstrated effects and possible harm. Insofar as the selection of upper limits tended to favor technological development at the expense of maximum personnel protection, the U.S. standards set by the military and ANSI were industry based. By contrast the willingness of Soviet and East European scientists to use subjective data, interpreting as hazardous anything that leads to discomfort, to accept cumulative models, and to dispense with complete scientific rigor pushed their standards toward lower limits—to the point nearest to zero risk. Insofar as the selection of lower limits could have retarded economic development in favor of maximum personnel protection, the 10μW/cm² standard can be looked at as a workers' standard.

Assessing the Validity of Standards

The standards set in the 1950s and 1960s in the United States and the USSR did not end the microwave debate. RF standards have remained controversial and still account for the majority of research and discussion on RF bioeffects. Supporters of the 10mW/cm² (or its recent replacement and near equivalent, ANSI C95.1-1982) still believe that science supports the view that this level of exposure can be considered safe, within the limits of present scientific understanding. Critics of the

10mW/cm^2 standard believe that it does not provide adequate protection.

The fact that military and ANSI standards did not end the microwave debate is not surprising. Since standards are not and cannot be the product of science alone—of necessity they are in part subjective—no standard can be right or wrong. It is possible to argue from one point of view that both the U.S. and the USSR standards are right. Insofar as both standards meet the objectives of their respective policymakers, both could be internally consistent and, in this sense at least, correct. But when the U.S. and USSR standards are assessed for internal consistency, both fall short of being valid or correct.

It is difficult for Western writers to assess the full meaning of the lower Soviet-bloc standards. We do not know how and to what extent they have been applied and how rigorously they are enforced, especially in the military context. Many Westerners believe that the 10μW/cm^2 is enforced in the East only to the extent that it is practical to do so. Questions have also been raised about the Soviet-bloc attitude toward science. While willing to dispense with the rigors of science at times, Soviet-bloc officials have relied heavily on science in setting standards. A great deal of their thinking has been moved by the acceptance of reported low-level effects. If this is so, then Soviet-bloc scientists would seem to have the same burden of scientific proof as is required in the West. It is contradictory to rely on science and to accept science that is defective. Defective science is not science and therefore cannot be used in scientific deliberations.

For their part, U.S. scientists and policymakers, consciously or unconsciously, have consistently misrepresented the philosophical underpinnings of the 10mW/cm^2 standard. Supporters of the standard like to convey the impression that this figure represents an objective extrapolation from scientific fact. Although science undoubtedly played an important role in the evolution of the 10mW/cm^2 standard, at every stage in its development it was significantly influenced by subjective factors.[53] The dominance of the thermal model was the product of bureaucratic decisions that vested control of standard setting in the hands of a few key individuals, notably Knauf, Schwan, and their successors. Subjective judgments on tolerance, margin of error, safety, and risk were made on every standard, primarily by persons heavily committed to thermal thinking and without serious discussions of the values inherent in these terms. Decisions about the burden of proof were made on personal and social grounds. U.S. standard setters subjectively decided to ignore missing information when discounting hazards but not when rejecting athermal effects. Again and again arbitrary, subjective judgments were made along the road to the 10mW/cm^2 standard. This fact cannot be denied.

It was therefore incorrect and deceptive to have argued, and is incorrect to continue to argue, that 10mW/cm^2 or figures that approximate this number must be accepted because they represent an objective extrapolation from scientific fact. Although policymakers attempted to be objective and scientific during the early years of standard setting, they did not live up to their reported ideals. When they drew up standards, they ignored philosophies and values. When they promulgated standards, they forgot that they had ignored philosophies and values. The extent to which these generalizations are true can perhaps best be seen by turning to scientific data that did not fit with the thermal point of view and examining how scientists reacted to these data when major policy decisions were made.

4

The Athermal Dilemma

It seems likely that neural function, and therefore behavior, are indeed disturbed by low intensity microwaves. . . . The behavioral studies for the most part leave much to be desired in the way of experimental control and clearly specified qualitative results. Nonetheless, the studies consistently and repeatedly report that human beings do exhibit behavioral disturbances when subjected to low intensity microwaves.—R. J. MacGregor, Rand Corporation, Santa Monica, California, September 1970

An athermal or nonthermal effect is one that is not thermal; it is an effect that seems to appear without heating. Determining the cause of such effects has intensified and polarized the microwave debate.

The demand for causal explanation is an important part of the scientific process. It provides a basic tool for separating fact from fiction. Scientific truth depends not only on observation but on causal explanation. Until the occurrence of a physical phenomenon can be explained, until some causal sequence is identified, it is difficult to be sure that what is being observed is in fact a genuine and discrete physical phenomenon. Causal explanation is critical to rigorous scientific explanation.

There is more to causal explanation than simply scientific rigor. Scientists accept as facts phenomena that cannot be explained. Newton's law of universal gravitation was accepted by many generations of scientists even though they could not explain how it operated in nature. Physicians routinely accept the existence of diseases that have no known causes. Acceptance or rejection on the basis of causal thinking is a selective process, one that brings in many factors besides the internal advance of science as an intellectual discipline. Such factors have frequently played a crucial role in shaping the development of thinking on so-called athermal effects.

Athermal Effects and the Medical Profession

The first major debates over the possible existence of athermal effects began in the late 1920s. Technological innovations developed at this time, principally the vacuum tube, made it possible to produce higher-frequency, "shortwave" energy that was uniform in character and propagated at discrete lengths. It was only a matter of time before researchers began to explore the biological effects of the new shortwave current. Its effects were soon being compared with those of the longer wave current that had been used in all previous bioeffects studies and comprised the principal tool for diathermy treatment.

One of the first scientists in the United States to study the bioeffects of the new shortwave technology was J. W. Schereschewsky, a surgeon employed by the U.S. Public Health Service and later an associate in preventative medicine and hygiene at the Harvard Medical School. Schereschewsky placed mice between two metal condenser plates, which were attached to a vacuum tube, and determined how long they survived when the power was turned on. As he tested the effect of different frequencies, he found that the lethal-dose time varied considerably (figure 4). At the lower end of his test range (8.3 MHz, roughly the lower end of the AM band), it took nearly twenty-five minutes to kill a mouse. At slightly higher frequencies (20MHz), the lethal-dose time dropped dramatically to about five minutes and then gradually rose again as frequencies approached 140MHz (the shortwave radio range) to over twenty-five minutes. These results left no doubt in Schereschewsky's mind that RF bioeffects were frequency dependent.

A number of things about the differential death times perplexed Schereschewsky. He could not immediately account for the shape of the curve—steep at one end, gradual at the other. If a single effect were being manifest, the relationship should have been constant. Different slopes suggested that different mechanisms might be involved. He knew that one of these mechanisms was heating. When he felt the mice immediately after death and took rectal temperatures, he discovered that most had experienced a body-temperature rise of 5° to 6°C.

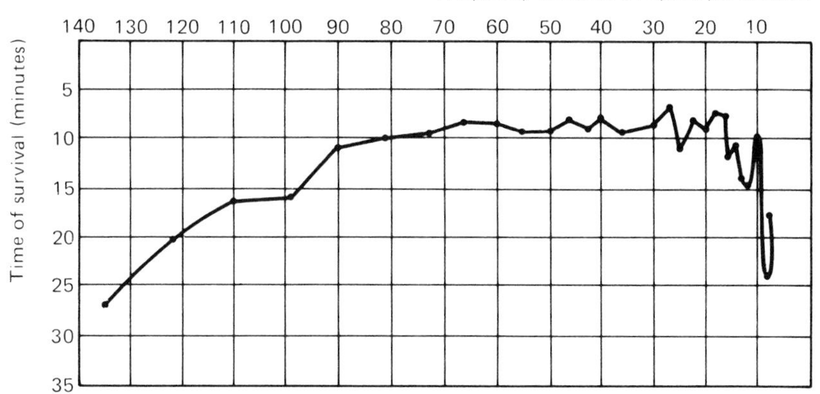

Figure 4
Survival times of mice subjected to currents of various frequencies. From J. W. Schereschewsky, "The Physiological Effects of Currents of Very High Frequency," *Public Health Reports* 41 (September 1926): 1958.

Clearly many of the animals were dying from overheating, but this was not uniformly the case. One mouse that succumbed at the expected time did not feel warm at death and had a body temperature of only 39.2°C, a value very near the normal body temperature for a mouse—37° to 39°C. Accordingly Schereschewsky was willing to argue only that "part . . . of the symptoms is due to heat retention," which left the door open for other symptoms not produced by heat.[1]

From his experiments on frequency dependency, Schereschewsky turned to tumor studies. He reasoned that if RF-energy–tissue interactions were specific, then perhaps frequencies could be found that would destroy only one type of tissue. The utility of the selective destruction of cancerous tissue made it an obvious target for investigation. Again, his test animals were mice, which had tumors transplanted onto their bellies. Once the tumors became established, they were treated by pinching them between small condenser plates attached to a shortwave generator. The results of the selective irradiation were dramatic. Of 400 mice with tumors that were treated, only 22 (5.5 percent) eventually died of tumor-related causes. In nearly 100 (25 percent) of the mice, the treated tumors quickly shrank in size and disappeared, leaving the mice to live normal lives and die of other causes.

Schereschewsky again cautiously turned to factors other than heat to explain why the tumors disappeared. Immediately after treatment the tumors did not feel warm to the hand. Schereschewsky therefore concluded that there was no "significant heating of the [tumor] tissues" and that heat was not the primary cause of remission. Some mechanism other than heating seemed to be involved. "The action of the electrostatic field in which the tumors are placed for treatment" did not "seem to be the same as that in medical diathermy."[2]

Reaction to Schereschewsky's work was immediate and negative. Most scientists were comfortable with the vague connection of all effects to a known mechanism, heating. They could not accept the specific attribution of some effects to unknown mechanisms. One typical negative response came from a biochemist at the Rockefeller Institute for Medical Research in New York City, Ronald V. Christie.

Christie began his bioeffects work by studying the response of individual cells to the long waves used in diathermy treatment. His objective was to discover whether all of the RF energy absorbed by a cell was converted to heat. His experiments showed that it was not. When he balanced input energy with heat production, there were discrepancies. A small percentage of the response of living cells "to the diathermy current seems not to lead to the production of heat."[3] However, these results did not convince Christie that there were athermal effects. In an article that appeared one year after his cell study, he and his colleague Alfred L. Loomis reviewed Schereschewsky's evidence and concluded that all of the effects could have been explained as thermal. Moreover they objected to the fact that Schereschewsky had not taken care in ruling out heat effects: "Schereschewsky states that the tumors did not feel hot, but on the other hand the microscopic picture of the tumors after radiation suggested coagulation necrosis [a heat effect]. It should be borne in mind that any heat developed in the tumor would rapidly be disseminated after the current was turned off. Direct temperature measurement in the substance of the tumor during radiation would be necessary to rule out the probability that the structural changes observed were not the result of heat." Christie and Loomis wanted proof that there was no heat production. In their opinion "the burden of proof still lies on those who claim any biological action of high-frequency currents other than heating."[4]

In placing the burden of proof on the athermal camp, Christie and Loomis began an era of double standards that has persisted in the RF bioeffects field to the present day: they demanded of their scientific opponents standards that they themselves did not meet. Schereschewsky had to prove beyond a shadow of a doubt that the effects he was observing took place in the complete absence of any heat production. But Christie and Loomis believed they did not have to prove with equal certainty that "the effects produced in animals can be fully explained on the basis of the heat generated by high-frequency currents which are induced in them."[5] They and their many followers were content with simple parallelism: if heat and effects were found together, it was assumed that the two were causally connected and represented the complete bioeffects story; however, parallelism and general observation were not accepted for verifying athermal effects.

Schereschewsky was not able to meet the burden of proof. When he followed the suggestion advanced by Christie and Loomis and measured the internal temperature of tumors during treatment, he found that heat had been generated. In some cases the internal temperature of the tumors rose as much as 10°C in less than two minutes. He also found that he could destroy the tumors by heating them in other ways to temperatures comparable to those achieved with RF heating.

The results of these follow-up studies, which were never published, tentatively moved Schereschewsky into the thermal camp. In an address to the Radiological Society of North America in December 1931, he concluded that "it is evident that the curative effects noted [in the tumor experiments] were due to the heating of the tumor cells in the high-frequency field." As for other athermal experiments, Schereschewsky simply summarized their content and noted that heat could account for some of the effects observed. Since it was almost always impossible to rule out any heat effects, the athermal case could not be proven. "Such," he concluded, was the status of "present knowledge in regard to the biologic action of ultra-high frequency electromagnetic waves."[6]

Despite the reservations of Schereschewsky and others, investigations into possible athermal effects of the new shortwave radiation did continue. A research team at the Western Pennsylvania Hospital in Pittsburgh reported being able to weaken (attenuate) the effect of disease toxins by irradiating them with 158 MHz (shortwave) RF radiation. During the experiments the temperature of the toxins was carefully controlled to avoid heating. The attenuation effects appeared even under the controlled conditions. Although "the exact nature of the mechanism of the change" was not known, the tests did suggest "that the action occurs without [any] heat effect."[7] But follow-up studies failed to uncover additional evidence to support the initial findings, prompting the research team that discovered the attenuation effect to downplay its significance. "The meager character" of the earlier athermal findings (the toxin attenuation studies) did not seem significant enough to be considered "in the interpretation of clinical or other studies of high frequency currents."[8] For clinical purposes, only thermal effects seemed important.

It was, however, from the clinic that the most important evidence for athermal effects began to emerge over the course of the 1930s. Many physicians who regularly used diathermy to treat patients experimented with the new shortwave equipment when it became available. Some of those who tried the new equipment were impressed by the results. A German researcher, Erwin Schliephaki, after studying the effects of RF radiation on small animals, decided to see whether shortwave treatment could cure a painful sore he had on his nose. The cure worked, and Schliephake immediately began using "shortwave diathermy," as it soon was dubbed, on patients. Within a few years major books were being written in Germany on shortwave treatment and its benefits. By the mid-1930s many U.S. physicians had followed the lead of their German colleagues, spurred on by an aggressive advertising campaign (figure 5).[9]

The rapid adoption of shortwave diathermy techniques by growing numbers of physicians in the United States alarmed the medical profession. Claims about the effectiveness of the new equipment seemed to be getting out of hand: "The medical profession has been extensively circularized with hyperenthusiastic literature extolling the advantages of this new form of therapy. Extravagant therapeutic claims have been put forward, based largely on the writings of Erwin Schliephaki, a German physician, whose assertions have been not only unconfirmed in this country but also partially refuted on the continent."[10] To check the flow of these extravagant claims and to avoid the impression that "this form of therapy has unusual healing powers and other clinical advantages," the Council on

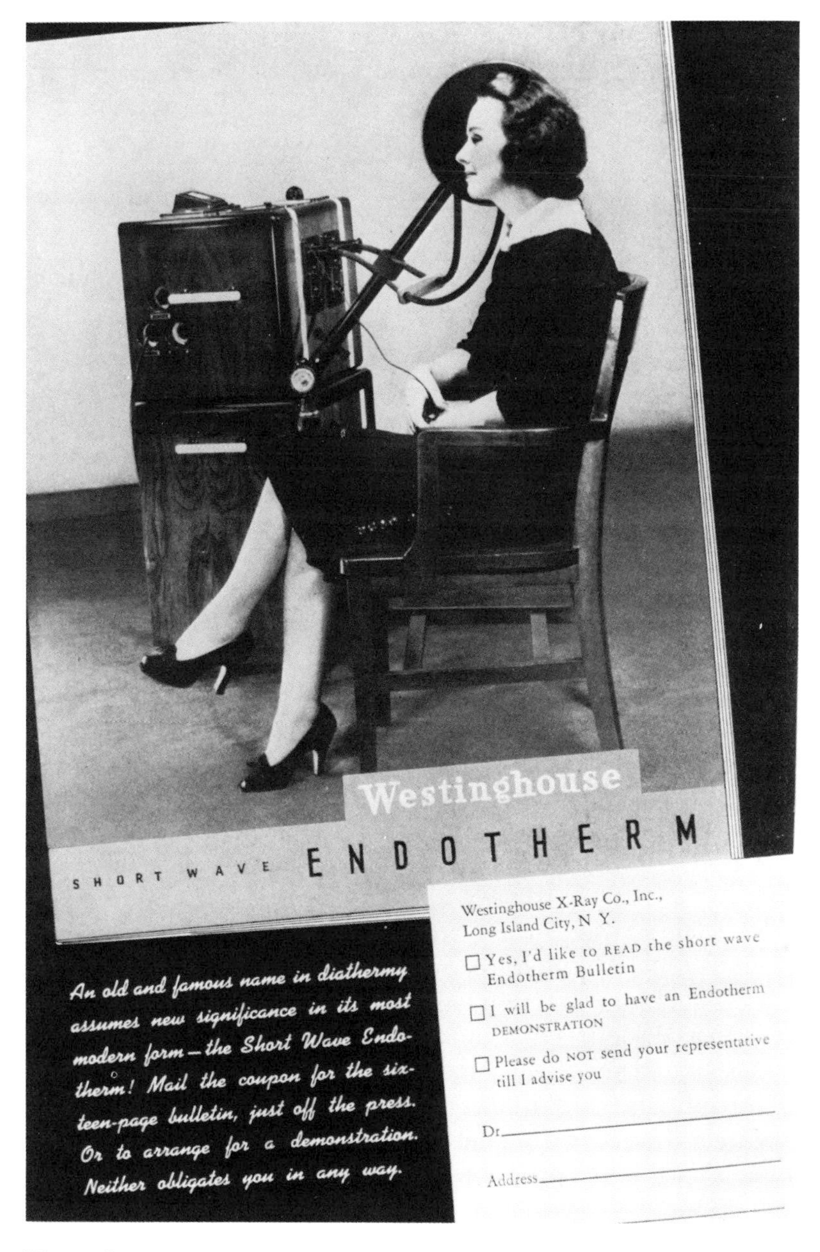

Figure 5
Advertisement for a shortwave diathermy machine. From the *Archives of Physical Therapy* (1938).

Physical Therapy (CPT) of the American Medical Association ordered an official investigation into the diathermy issue.[11]

The entry of CPT into the microwave debate added a new dimension. Early disagreements about the possibility of athermal effects remained mostly a matter for internal discussion between a limited number of concerned scientists. The CPT review, by contrast, had a much broader audience. It had bearing on major segments of the medical profession. The conclusions published under the sponsorship of the CPT had the potential for determining how an entire profession conducted its affairs. For the first time scientific judgments became linked to major social consequences; science and economics became, and would hereafter remain, intertwined.

The options CPT reviewers faced were these. If their report accepted shortwave diathermy as superior to traditional diathermy treatment for reasons that were not fully understood and perhaps not totally related to thermal effects, the door could be opened to widespread and uncontrolled adoption of the new equipment. This in turn could weaken the influence that organizations such as the American Medical Association exerted over medical practice through science. A conservative report that stressed science would have just the opposite effect. If scientific doubts were used to restrict application, the number of practitioners and accepted therapeutic practices would be limited, thereby keeping a tight rein on professional activities.

The review article on diathermy ultimately sanctioned by CPT in April 1935 took a hard line on the RF bioeffects issue. All "specific biologic action," the term used to refer to athermal effects, was discounted because of a lack of conclusive evidence. Any specific effect that seemed at all possible was attributed to heat production. No doubts were resolved in favor of possible athermal effects. The burden of proof remained "on those who claim any biologic action of these currents other than heat production."[12]

The total rejection of athermal effects was accompanied by another restriction. Some proponents of shortwave diathermy were convinced that the new equipment provided a more efficient means for delivering uniform heat to body areas beneath the skin surface. This claim was also rejected. CPT's researchers had conducted their own studies of heat penetration and found "a thermal gradient from the hot skin to the less hot tissues within," not a buildup of temperature below the skin surface as some maintained. They also could find "no evidence . . . that shortwave diathermy" produced "a more uniform penetration of heat into the body than the conventional diathermy."[13] In sum there seemed to be no evidence for unusual claims of any sort. Diathermy simply provided a useful tool for delivering heat to the surface of the body. This was the only therapeutic application sanctioned by CPT and, by implication, the only biological effect that passed the muster of scientific accountability.

CPT's report did not end debate, however. One medical practitioner in New York argued at length that it was absurd to restrain the development of shortwave diathermy treatment simply because it might be dangerous if used incorrectly—this was one excuse advanced to launch the initial CPT investigation—or because its mode of action could not be explained. He countered by noting that "half the cases of malpractice in New York State are due to burns caused" by traditional diathermy treatment. Thus misuse was already a problem, and there seemed to be no reason why the situation would be worsened by turning increasingly to shortwave diathermy. As for the lack of biological knowledge, the New York practitioner pointed out "that a great many therapeutic procedures are biologically far less completely explained than short waves, . . . yet, they are advocated and applied." Thus if standards were applied consistently—which they were not under the double-standard principle—there was no reason why shortwave diathermy should not have been given the benefit of the doubt.[14]

This practical, see-what-happens point of view remained the unofficial position of many doctors. If the treatment worked, they felt free to use it and did. Throughout the remainder of the 1930s and well into the 1940s, numerous articles on shortwave diathermy appeared in which treatments were described for disorders as varied as intestinal upsets, cancer, and arthritis. But this was not the official point of view, the view that was applied when decisions had to be made on acceptability or control. When it came to taking official action, the research-oriented physicians who supplied advice clung firmly to the position that unless indisputable scientific evidence were found to the contrary, there were no athermal effects. It is this official and professionally motivated opinion that military planners took up when they made decisions about the bioeffects problem in the 1940s and 1950s.

Athermal Effects and the Military

The lag in diathermy research that came with the outbreak of World War II brought to an end the thermal-athermal debate in medical circles. When physicians started to renew their diathermy studies after the war, new problems emerged that delayed acceptance of microwave diathermy. By the late 1940s data had been collected that suggested that shortwave treatment did produce a differential heating effect, earlier CPT claims notwithstanding. Several investigators were able to raise internal temperatures without significantly altering skin temperatures. Since the body's main heat sensors are located in the skin, this effect raised concerns. If the body could be heated or overheated internally without any external sensation of pain, a genuine danger did exist. Before microwave diathermy could be approved for general use, this issue had to be resolved.[15]

Uncertainty over the safety of microwave diathermy was further complicated in the postwar years by the discovery of RF-produced cataracts in rabbits and damage to the testicles of irradiated rats. Now, in addition to the straightforward chance of accidental damage, there seemed to be the possibility of unwanted side effects. The cataract issue was troublesome because it raised the possibility of delayed effects. The fact that the Iowa team had discovered that cataracts could take as long as forty-two days to develop meant that short-term effects were no longer all that counted.[16] This in turn raised the possibility that the immediate response experienced in a physician's office was not all that needed to be considered in assessing the potential dangers of diathermy treatment.

The testicular degeneration studies cast more doubt on the thermal solution. Again it was the team of researchers at the University of Iowa, working under contracts from Collins Radio in nearby Cedar Rapids, Iowa, that prompted additional thinking about athermal effects: "The outcome of this experiment [with rats] clearly shows that testicular damage will result from 12 cm [2450 MHz] irradiations of a temperature below that of the abdominal cavity and below that necessary to cause injury by infrared exposure." The absence of any noticeable temperature rise unquestionably raised the possibility of athermal effects; it suggested that "damages may result in part from factors other than heat."[17]

The medical community dealt with these uncertainties by abandoning diathermy. By the mid-1950s papers on diathermy were becoming fewer in number, and the field in general was being abandoned. (In recent years there has been renewed interest in exploring the potential of diathermy, especially for treating cancer.) The scientists beginning to pursue the bioeffects problem for the military did not have this luxury; they could not turn their backs on these new problems. By the late 1940s radar had become indispensable to the military. Therefore the problem of side effects had to be addressed, which meant attempting to come to grips with the issue of athermal effects.

As the question of athermal effects was taken up in the years leading to and during the Tri-Service era, the critical attitude adopted by physicians in the 1930s once again prevailed. Before any thought would be given to athermal effects, there had to be firm evidence indicating that such effects did exist; otherwise the prevailing sentiment automatically sided with thermal thinking. A Ph.D. dissertation written in 1951 concluded, "The consensus of opinion after many experiments designed to prove or disprove the theory of specific or athermic effects of h-f [high-frequency] radiation is that all observed physical and pathological changes in biological tissues can be explained by the effect of heat alone."[18]

This conclusion had in fact not been proved. The athermal school was not the only one that lacked evidence. Those who supported the thermal solution also found it "difficult, in many instances, to rule out completely the possibility of athermal effects."[19] The burden-of-proof argument could cut both ways, but it was not applied both ways. Once again the double standard came into play. Throughout the 1950s and into the 1960s the most rigorous demands for evidence were made of those persons who claimed to have discovered athermal effects.

The overwhelming community commitment to thermal thinking severely limited the creativity of RF bioeffects research. Rather than attempting to learn from reports of athermal effects, the RF bioeffects community by and large devoted most of its attention to clarifying and proving what it already knew or to disproving claims believed to be false. This approach to research encouraged a single-mindedness that rigidly adhered to the thermal solution, a single-mindedness that can be seen in responses formulated when athermal effects were reported.

By far the most widely discussed athermal effect, both before and during the early Tri-Service program, was testicular degeneration. The articles that appeared describing this effect clearly stated that it occurred at power levels that seemed unlikely to cause much, if any, heating. It had also been shown to occur at power levels well within the range of existing radar equipment, as a confidential 1955 air force survey demonstrated.[20] The reported existence of this effect could not be doubted. But other things about it could. The 1955 air force survey, for example, placed a new burden of proof on the athermal argument—clinical importance. True, the effect might exist, perhaps in the absence of heat, but did it have any bearing on the health of personnel being exposed to radar? The air force's reasoning suggested that it did not: "The production of similar effects by mild exposure to infrared . . . suggests that this is a common thermal effect and may be of no greater significance than the reaction experienced by a man in a hot bath."[21]

This rejection notwithstanding, other researchers did attempt to assemble more data. One research team organized by Stephen Ficker at MIT looked for testicular effects in rabbits exposed to a 10–60 mW/cm^2 RF source. They found no significant differences between the testicles of control and test animals and concluded that there were no RF-related testicular effects. Those effects that they did discover—and report at the first Tri-Service conference—such as death in 30 minutes at 60mW/cm^2, were all thought to be caused by heat.[22] Ultimately, general agreement was reached that although "temperatures are not necessarily the prime factor in a number of the conditions which have been produced, . . . most of these shown in these meetings [such as testicular degeneration] are the result of excessive heat."[23]

The relevance of the MIT studies to the testicular degeneration question was slight. It was well known at the time that rabbits are more heat sensitive than rats. Therefore using rabbits to check conclusions about athermal effects reported in rats made no sense. Moreover it is clear from the brief report of the MIT work that thermal stress had a major effect on the test animals. Exposure levels as low as 30mW/cm^2 killed the rabbits, presumably because of overheating. Given the obvious thermal stress and the difference between rabbits and rats, the fact that athermal testicular degeneration could not be detected in rabbits had no scientific bearing on the athermal question.

This line of reasoning was not allowed to develop under the pressures of thermal thinking. Replication was either not encouraged or ignored when it seemed to verify the original findings.[24] The issue of clinical importance was not pursued. Even though it was realized that "low morale . . . might prevail among those having potential microwave exposure, if . . . workers came to believe, rightly or wrongly, that permanent testicular damage could or would result" from exposure, extensive personnel studies were not conducted. Advisers at GE recommended that monitoring be undertaken, but it was not.[25] In sum, rather than leading to increased scientific activity, doubt and contradictory reports about athermal effects were used as excuses to ignore them. Throughout the 1950s scientists continued to study what they understood best, the thermal phenomenon, not what they did not understand at all.

Much the same response followed in the wake of reports on another general class of possible athermal effects, sensory and behavioral effects. As early as 1955 the air force reported that personnel experienced "a tinnitus or ringing of the ears . . . during exposure in the strongest cross section of" a radar beam. A year later a personnel survey conducted at Lockheed Aircraft Corporation in Burbank, California, noted "complaints of buzzing vibrations, pulsations and tickling about the head and ears . . . in association with S band, high powered antenna."[26] Both reports could have been interpreted as suggesting some type of central nervous system effect, thermal or athermal in origin, but no effort was made to find out the bearing such effects had on the RF bioeffects question and its thermal bias.

The research conducted during the Tri-Service era discovered other possible examples of central nervous system effects. Researchers of the University of Miami School of Medicine reported at the third Tri-Service conference that they could alter the behavior of rats by exposing them to as little as three daily 15 to 30 minute doses of moderate level (109mW/cm^2) RF radiation. They measured effects by placing the rats in a modified Skinner box after exposure and seeing how rapidly they performed specific tasks. The irradiated rats did not perform the tasks (such as depressing levers to get food) as rapidly as their control counterparts.[27]

More dramatic behavioral–central nervous system effects were described by Sven Bach, a researcher at the U.S. National Institute for Neurological Diseases. Bach's group restrained their test animals, rhesus monkeys, with plastic clamps and irradiated only the head region with carefully controlled, moderate-level (about 64mW/cm^2) RF radiation. The effects seen under these exposure conditions were dramatic. When exposed, the animals went through cyclical periods of drowsiness and arousal. During the drowsy periods they stared with a fixed gaze and were "unresponsive to touch, pain, light, and sometimes to sound stimuli." Then they moved into a period of arousal characterized by rapid side-to-side head movements. "By alternately switching the transmitter on and off, one of these animals was brought to the point of successive arousal and complete relaxation, in a 20-second cycle, reacting like a puppet on the end of a string." Under proper exposure conditions convulsions could be induced and death brought about in as short a time as 2 minutes and 55 seconds.[28]

Follow-up studies did not point to heating as the cause of these effects. Pathological studies turned up changes that seemed to be the results of direct electrical stimulation. Bach and his team would not speculate whether these findings pointed to thermal or athermal mechanisms as the primary agent, but the results were worth further exploration.

The monkey experiments prompted some public interest. In March 1959 Pearce Bailey, director of the National Institute of Neurological Diseases, related more details on the experiments to a House Appropriations Subcommittee. He speculated, according to a report in *Aviation Week,* that similar central nervous system effects might account for "mysterious airplane accidents."[29] Such speculations were picked up by other journalists, and for a few months there was a great deal of talk about the significance of the tests and what ought to be done. But again little effort was made to continue these studies, even though a formal comment on Bach's paper at the third Tri-Service conference strongly recommended further investigation: "Clearly the work will have to be continued and extended in order to delimit the conditions under which these phenomena are produced and to obtain quantitative relations between dose and effect. Statements at the present time as to the mechanism of these effects would be premature. The possibility should not be overlooked that the intense electric fields which occur may have some sort of direct action on nerve tissue."[30]

Further study or even replication did not follow. In work conducted at Tulane University for the specific purpose of testing the athermal hypothesis advanced by Bach and others, not one of Bach's experimental procedures was duplicated. The test animals were decerebrated cats, not live monkeys; 100MHz pulsed radiation was used instead of the 38MHz modulated CW radiation that Bach had found to have maximum effect; and thermal doses (200mW/cm^2) replaced the potential athermal doses (64mW/cm^2) of the original monkey tests. Despite such differences the Tulane group confidently announced that "it is believed . . . that thermal stimulation of afferent sensory pathways of peripheral nerves accounts fully for the behavior response to microwave irradiation shown by experimental animals."[31] This conclusion was sufficient to convince those who wanted to be convinced that the thermal hypothesis held and that no further work needed to be done.

A similar response greeted the one researcher, Allan Frey, who took the reports of direct auditory stimulation (microwave hearing) seriously. While working at the GE Advanced Electronics Center at Cornell University, Frey came in contact with a GE technician whose job was to measure radar energy levels at field installations. This technician claimed that he could hear radar. The report intrigued Frey, so he set about designing a few simple scientific experiments to test its validity. Subjects were blindfolded, situated far enough away (up to 100 feet) from radar units so they could in no way hear any normal noise the units made, and were subjected to both regular and random irradiation. When correlations were run between actual and perceived exposure under these conditions, the results left no doubt that the test subjects could sense when they were being subjected to microwave radiation. Even some deaf people could hear radar, and at very low power levels, as low as 0.065mW/cm^2.[32]

When Frey reported these results at an Aerospace Medical Association meeting in April 1961, skepticism prevailed. A few people for years had claimed to be able to hear radio programs through fillings in their teeth and to communicate with Martians. Even if the reports were true, these experiences seemed to pose no apparent threat to health and so were dismissed. A serious scientific dialogue on microwave hearing did not develop until the early 1970s. Even so by the time Frey published his findings in late 1961, conditions were changing. A few cracks were beginning to develop in the thermal facade, cracks that would in time bring the entire structure crashing down.

Soviet Research and Unexplained Effects

U.S. scientists learned in the late 1950s that their counterparts in the Soviet Union had been vigorously pursuing research on athermal effects for a number of years. Colonel Knauf talked with Soviet scientists in the summer of 1960 during a trip to Europe to attend NATO meetings and the Third International Congress on Medical Electronics. His contacts led him to conclude that the United States and the USSR were in substantial agreement on the RF bioeffects question. Knauf reported this conclusion to researchers assembled in New York City in August 1960 for the fourth Tri-Service conference and went on to endorse a suggestion made by the USSR Academy of Medical Science for "an interchange of papers on the subject."[33] As U.S. scientists learned more about the Soviet work, they discovered that Knauf's report to the fourth Tri-Service conference was far from accurate.

One year after the conference and Knauf's report, Soviet scientists traveled to the United States to attend the Fourth International Conference on Medical Electronics. The papers they presented plainly indicated that Soviet researchers, unlike their U.S. counterparts, were extremely interested in the consequences of low-level, athermal effects, particularly on the central nervous system. Work reported by Z. V. Gordon of the Academy of Medical Science attributed the cause of reduced irritability in rats after RF exposure to "changes in the neural cells, in receptor fibers, in various receptor zones, and especially in axodendritic and axosomatic interneuronal junctions in the cortex of the brain." Her colleague, A. N. Obrosov, noted that Soviet scientists had been studying personnel effects since 1942. The symptoms reported by radar workers led Soviet scientists to explore the possibility of using RF radiation in medicine. Their studies concluded that RF treatment "appears as a new, extremely promising and effective therapeutic method."[34]

Few U.S. scientists took these early reports seriously. The details of the experiments underlying the Soviet reports were usually sketchy, making it impossible to attempt replication or to check for errors. In addition U.S. scientists did not hold Soviet research in very high regard. Still the fact that the Soviets were investigating athermal effects on the central nervous system could not be ignored. If there were important discoveries to be made, the United States could not be caught napping, as it had been with the launch of Sputnik in October 1957. At the very least scientists had to keep an eye on Soviet developments in fields that could be related to space exploration.

The first serious efforts to learn more about the Soviet work date from shortly after the summer 1961 Medical Electronics conference in New York. Translators working for the National Aeronautics and Space Administration, the air force, the army, and Bell Labs in Whippany, New Jersey, began pouring through Soviet journals to find out just how much they knew about RF bioeffects. Over the next few years dozens of Soviet and East European scientific publications were rendered into clumsy English and distributed to a select group of government and military agencies. The picture these translations painted of the Soviet work contrasted sharply with the view Knauf had presented in 1960.

A brief 1962 report, "The Effect of an Electromagnetic Field on the Central Nervous System," by Yu Kholodov (translated in August 1962) openly and explicitly contrasted the Soviets' interest in low-level effects (below 10mW/cm^2) with the U.S. concentration on higher power levels (tens of watts/cm^2). Kholodov noted with some amusement that one U.S. experiment, Bach's selective irradiation of rhesus monkeys, "caused a big stir in the American press and even was discussed in one of the commissions of Congress." The Soviets dismissed such experiments as trivial; "the described influence . . . on a monkey can be compared with the influence of a high voltage current on a human being, which, of course, is fatal. But from this it does not at all follow that the current of a battery of a flashlight is dangerous for life."[35]

An article by A. S. Pressman translated at roughly the same time reported that pulsed radiation affected the heartbeat of rabbits at power levels as low as 3–5 mW/cm.^{2} Pressman also referred to an earlier report that found similar results using slightly higher doses (7–12 mW/cm^2) of CW RF radiation.[36] Translations of other reports described the results of Soviet surveys of radar workers and the health effects they were experiencing. And perhaps most significant, a July 1962 summary of a major article published in 1960 correctly put on record the fact that the Soviets had concluded from all of this work that safety standards for workers had to be set well below the level of $10mW/cm^2$ that was then being adopted in the United States. Soviet workers were permitted to be exposed to only one-one-thousandth of the U.S. standard, $0.01mW/cm^2$, over a twenty-four-hour period.[37]

As significant as the data contained in these translations might seem, they were not widely publicized. Most U.S. scientists learned about the Soviet work only later in the 1960s and did not take it seriously until the early 1970s. The gradual influx of information about the low-level studies did not lead to reactivation of the Tri-Service effort or result in increased grant support for university research. But it did, when combined with the continued nagging problem of possible central nervous system effects being reported in the United States, keep a few U.S. policymakers aware of the athermal effects problem. Occasionally they tried to decide what, if anything, ought to be done. So it was that the Air Force Systems Command, a primary recipient of most of the translated Soviet reports, contacted the head of the Brain Research Institute at the University of California, Los Angeles, for the purpose of calling a conference on Neurological Responses to External Electromagnetic Stimuli. The conference took place on July 11, 1963.[38]

The announced purpose of the UCLA meeting was to discuss scientific explanations that might account for a number of reports of apparent central nervous system and behavioral effects being received by the air force. In one case the wife of an air force researcher in New York experienced "an unpleasant tingling sensation over large areas of her skin" whenever a recently redone bedroom electrical system was turned on. A report of a woman in California who could hear low-frequency radiowaves came from an expert well versed in electrical engineering and biomedical effects. A third case reported in some detail concerned the fate of an upstate New York family, the Binkowskis, who were driven from their home by mysterious noises. The probable cause of the noises was RF radiation from a local radio station, although no mechanism could be given for a cause-effect relationship. This "growing body of reports and observations that animals, including men, respond at least under some conditions to . . . electromagnetic radiation as if their nervous systems were involved" was used to launch a free-ranging discussion of the ways in which RF radiation could or could not affect the central nervous system.[39]

No consensus emerged from the UCLA meeting. The science and phenomena under investigation were too speculative to allow reasoned progress to be made. There simply were not enough hard data available to permit any sure conclusions. At the time the only objective evidence U.S. researchers had on possible direct central nervous system effects was Allan Frey's work on RF hearing, but conference participants could not agree on the significance of his work. Some questioned his experimental techniques, others drew different conclusions, and still others explained the hearing phenomenon as a mechanical or electrical phenomenon. About the only generalization that was not challenged was the suggestion that RF hearing could be further studied.[40]

When the data became less controlled, the problems of interpretation became even more troublesome. There was, for example, the strange radio receiver invented by "a rather eccentric, mushroom-eating, mental telepathy, air-plane-glue sniffing type psychic researcher" from the University of Chicago, Henry J. Kuharis. Kuharis had invented a device that made it possible for one person in a room to hear music being played on a tape recorder but not amplified, while others in the room could not. All the person had to do was hold a half-dollar-sized piece of metal covered with cellophane tape "up against your face or your neck, or I think it will work in your belly-button," hook up a few wires, and the result was "beautiful music, beautiful fidelity. It works." Kuharis claimed his device worked because he had discovered a way to stimulate directly the facial or auditory nerves. His original reason for contacting the air force was to interest them in buying a radio that would be fastened in the mouth and tied directly into the nerves of the teeth. As the air force spokesman noted in describing Kuharis's radio, the simplest way to deal with him would have been "to tell him to go and soak his head. . . . We could go and buy a Japanese transistor radio and stick it in our mouth and do just as well."[41] But the solution was not as simple as this. There was just enough left unexplained with Kuharis's radio, in the RF hearing phenomenon, in the episodes of persons being affected by electric wires and radio stations, and in the Soviet literature on low-level RF bioeffects to call for a discussion of what should be done.

At the end of the discussion, the options open came down essentially to two: the reports of the various phenomena could be dismissed as unsubstantiated and the matter dropped, or a concerted effort could be undertaken to find out what was going on. One of the final speakers, Lewis Bach, strongly urged that a choice be made: "These reports have been going on for many, many years and I think we ought to fish or cut bait, in other words, we ought to proceed on likely clues and likely lines and push them very hard to see if there is something or if there is not something, one way or the other. We can't dangle forever in limbo." Bach had compelling reasons for continuing. The unorthodox work discussed at the conference provided "potential ways of getting a better or more diversified understanding of the function of the central nervous system." Science could benefit from this further study. Moreover Bach had philosophical reasons for going on. Since humans evolved in a world permeated by electric and magnetic forces, he regarded it as at least likely that "there may exist centrally located [RF] receptors" in the brain. Study along these lines was not without justification and might even be rewarding.[42]

Bach's recommendation to continue the search was partially adopted. Following the UCLA conference, the military, which controlled the RF bioeffects pursestrings and therefore made the major policy decisions, decided both to fish and cut bait. Publicly talk of athermal effects was downplayed. Open contracts were not awarded for athermal or central nervous system studies, and in fact efforts were even made to keep information about central nervous system research from circulating too widely. Privately, however, the military and the State Department began work to try to determine whether there was any factual basis for a belief in the direct effect of RF radiation on human behavior and whether perhaps the Soviets had gotten the jump in exploiting such effects for espionage and military purposes. The primary motivation for the work was a desire to find out the purpose of a beam of microwave radiation that was being directed at the U.S. embassy in Moscow.

Barriers to Athermal Thinking

The fact that U.S. scientists and policymakers submitted so readily to the tyranny of thermal thinking and did not initially take athermal, low-level and long-term effects seriously has caused major problems for the entire RF bioeffects field. The lack of concern exhibited for effects that would, if present, cause the most human harm—subtle effects experienced in the normal course of life by large segments of the population—seems indicative of a callousness that could have been motivated only by some conspiratorial design.

To prove that there was some conspiracy requires evidence that key individuals knew that they were taking actions that could not be justified on the basis of existing information. In the years preceding the Moscow embassy crisis, there is no solid evidence to support the contention that some persons knowingly disregarded information when deciding how the RF bioeffects field should develop. There is, moreover, no need to resort to premeditated conspiracy to explain what went on. Events prior to the 1960s can be explained in other ways that, though not as simple as a conspiracy theory, more nearly mirror the complex reality of life.

Paramount among the other factors that explain the thermal bias is the fact, quoted widely at the time, that no one was dying or even getting sick from RF exposure. For many persons raised during the war years and the ensuing era of support for technological development, this fact alone was sufficient to overcome all concerns. It may not have been true that no one was dying. There was at least one report of a radar death in the scientific literature.[43] It was also not true that death was a fair measure of hazards or that the absence of immediate effects meant that eveything was all right. But such caveats did not carry the day. The single case of a radar death was disputed, and this fact was sufficient to allow the incident to be ignored. As for other more subtle effects, consciousness had yet to be raised by society in general as to their possible significance. As a result most policymakers, particularly in the military, had no trouble believing in the thermal solution. Within the context of the times, it made sense.

A second factor that stood in the way of accepting athermal thinking was the conservatism of science itself. Scientists in general have never excelled at challenging the orthodoxy of established world views. This fact is often obscured by the importance assigned to a few revolutionary figures such as Copernicus and Einstein. In stressing the achievements of these few individuals, we tend to forget that they comprise a very small minority and that it took decades for their radical ideas to be accepted by other scientists.

Between 1930 and 1960 the RF bioeffects field had no such radicals. Its members worked diligently to articulate the accepted world view; they did not seek to challenge it and even worked hard to protect it from criticism. As a consequence when policymakers turned to the community of experts for scientific advice, the advice they received supported prevailing points of view. Insofar as the orthodox scientific view also supported the belief that no one should have been dying from exposure to RF radiation, the general observations made by policymakers and the views of the scientific community tended to reinforce one another.

A third barrier to athermal thinking can be found in the social consequences of accepting the unorthodox point of view. Athermal thinking undermined accepted mechanistic thinking. It focused on the unexplainable, not on established cause-effect relationships. As such it opened the door to schools of thought radically opposed to the prevailing orthodoxies. In the 1930s and early 1940s the acceptance of athermal effects implied embracing medical practices that bordered on quackery. By the early 1960s the major reports of athermal effects seemed to come from psychological misfits. To take up company with the athermal camp came to imply taking up company with such quacks, psychological deviates, and, eventually, the Soviets. It was not easy for anyone interested in succeeding in the orthodox world of science to take up such company. The social pressures against athermal thinking were formidable and have remained so.

When these and other less important factors are put together, the thermal tyranny comes down to the fact that it was easier for most persons to accept this view. Through the Tri-

Service era there were ample research funds to keep thermal work going. The problems inherent in the thermal solution were important enough to justify ongoing investigation. Heat effects were not completely understood. The evidence that challenged the orthodox position was not overwhelming. Athermal thinking could be and was at the time ignored. But then the Moscow signal was discovered, and for a select population, the RF bioeffects problem suddenly took on a dramatic new look.

5

The Moscow Embassy Crisis

This is an issue that involves some rather sensitive intelligence matters and, therefore, we have been reluctant to discuss it publicly while we are attempting to safeguard the health of our employees—a matter which we have substantially done—and while we are trying to negotiate a solution to this difficulty. If these negotiations should prove impossible to conclude, we will then of course have to be more explicit in our explanations. But while these negotiations are going on, I would prefer not to go further than to say . . . that our principal concern is the health and safety of our employees.—Secretary of State, Henry Kissinger, on the Moscow embassy crisis, March 6, 1976

With the termination of the Tri-Service Program the pace of RF bioeffects research in the United States declined. Some research did continue after the program ended, primarily under navy sponsorship; papers were published on potential hazards; and ANSI's standard committee drew up RF exposure guidelines by late 1966. But compared to events of the late 1950s the level of public activity declined noticeably in the 1960s. With the standards problem apparently solved, there was no urgency for continuing high-level support for RF bioeffects research. Behind this quiet facade, however, a new concern was growing as a result of the discovery that a beam of low-intensity microwave radiation was directed at the U.S. embassy in Moscow.

The origins of radiation problems in Moscow go back to 1952 when the U.S. embassy was moved from a site near the Kremlin to a newly renovated apartment building several miles away. Shortly after the move routine radiation checks turned up unexpected readings. One sweep made in advance of Vice-President Richard Nixon's trip to the Soviet Union in 1959 discovered high ionizing radiation levels in sections of the ambassador's apartments, including one of the rooms where Nixon was scheduled to sleep. Similar checks for nonionizing radiation conducted as early as 1953 detected the presence of a microwave signal apparently beamed at the embassy from a nearby building.

The Moscow signal, as it came to be known, at first did not appear on a regular enough basis to cause concern. But over time it became more constant. The State Department responded to this situation by shipping electronic equipment to Moscow to monitor the signal. The figures on power, frequency, and modulation collected during the early 1960s left little doubt that the Moscow signal was a carefully manufactured beam of radiation aimed at the embassy and its personnel. This much was obvious. What was not obvious was the reason the Soviets were beaming the signal. To find out, the State Department, with the help of the military, initiated a new and classified research project on low-level RF bioeffects.

Early Response to the Moscow Signal

It is impossible for persons outside the small coterie of officials who dealt with the embassy crisis to learn the purpose of the signal. In response to Freedom of Information Act requests, the State Depatment, the military, and the Central Intelligence Agency have made some documents available for public use, but national security restrictions have prevented these agencies from making the full record available.[1] The information that has been released is edited and incomplete. Several letters released to me in response to my requests were completely censored. Therefore the conclusions I make in this chapter are tentative.

The lack of written documentation can be overcome partially by turning to information released through interviews and public statements. These sources of information are useful for filling in details but still do not answer the basic question of motivation. Those who took part in the events of the late 1960s and early 1970s are reluctant to talk even in private about the reason(s) the Soviets began radiating the embassy. There appear to be official views that explain the episode and perhaps undisclosed documents in the archives that answer all questions, but such documentation has not been made public.

The State Department and the military eventually learned that they were dealing with a low-intensity (about 0.1–24μW/cm^2 in the embassy building), high-frequency (in the gigahertz range), modulated signal, but this information took time to assemble.[2] At first they knew only that the signal was there and was being directed at the embassy. They were not sure of its characteristics or intensity and apparently did not know its purpose.

With so much uncertainty the information contained in the translations of Soviet bioeffects research papers made in the early 1960s took on special significance. These papers provided one and perhaps the only real explanation for the Moscow signal: to affect the behavior or health of embassy personnel. The reasoning that led to this conclusion was inescapable and clearly spelled out in secret memos circulated at the time.

In one such memo, dated May 13, 1965, a physicist at the Advanced Research Projects Agency (ARPA), Sam Koslov, commented on the possible objective of the Moscow signal for the State Department's head of security, Charles Weiss.[3] At the time government sources erroneously estimated that the intensity of the signal was about 0.5–1mW/cm^2. (The lower μW/cm^2 figures were not reported until late 1967 or early 1968.) This figure, Koslov noted, "put the irradiation of the embassy . . . a factor of 100" over the Soviets' own RF safety standard. This fact, plus the directional and modulated character of the signal—it was not a random continuous signal such as would emanate from a radio broadcast station—led to the conclusion that the radiation was being deliberately beamed at U.S. personnel at levels that the Soviets themselves regarded as dangerous.

Koslov did not know why the Soviets were bombarding the embassy with modulated microwaves. He rated antipersonnel implications as possible but not probable: "A possible explanation of the Moscow signal may reside in an attempt to produce a relatively low-level neurophysiological condition among embassy personnel. Although the detailed studies of the signal do not give this a high probability of interest, nevertheless, the intensity of the signal relative to the published Soviet safety standards is sufficiently high to furnish a logical basis of protest on microwave radiation grounds." This possibility rested on the several hundred reports on RF bioeffects published by the Soviets. Since they apparently believed that "excessive fatigue, coordination loss, changes in blood pressure, in heart rate, biochemical changes in the blood, and some loss of sensory ability" could result from low-level RF exposure, producing such effects had to be regarded as at least a possible objective of the signal.

Koslov concurred with plans being developed at the time and suggested to Weiss that "a sober and systematic program of research" be planned to look into the problem. A straightforward experiment subjecting higher primates to a duplicate Moscow signal would accomplish this objective. If the primates experienced adverse effects, the United States would then have a "data base for possible use in a protest action." Although pursuing such research might raise suspicions, Koslov noted that "the counterintelligence problem can be easily buried in a program of routine irradiation at low-levels." The military had pursued research on bioeffects before; there was no reason why they could not do so again even if the motivation for it had changed.

At the time Koslov wrote this memo, plans were being made to pursue not only the proposed primate study but one other project, a survey of the health of embassy personnel. The directive making these two studies the focus of official action came from the highest levels of government. Exactly how and when the heads of government decided to take action is not revealed in the documents made public so far, but it is possible to piece together where within the government key decisions were made.

Once the deliberateness of the USSR's actions became apparent to the U.S. intelligence community, the embassy problem passed quickly up through official ranks. Secretary of State Dean Rusk became involved during the early data-gathering stage when U.S. officials were trying to find out if the signal was being aimed specifically at the embassy.[4] Most of the data gathering was done by the CIA. Thus from the early 1960s on, at least two agencies in the U.S. intelligence community were aware of the problem. Probably some segments of the military community also played an early role.

From the individual agencies the embassy problem passed to the central oversight board for all intelligence, the United States Intelligence Board (USIB). This board and its chairman, the director of central intelligence, has the job of interpreting the intelligence reports of its member agencies and plotting a uniform course of action. Both the State Department and CIA were member agencies of the USIB at the time, along with the Defense Intelligence Agency, the National Security Agency, the Joint Staff, the Atomic Energy Commission, and the Federal Bureau of Investigation. By late 1964 or early 1965 the USIB had assigned the embassy problem to its Technical Surveillance Countermeasures Committee (TSCC). On April 16, 1965, TSCC met to review all data and issued directives for a two-pronged investigation: a health survey and an investigative research program such as the primate study. All of these actions ultimately were under the direction of the White House.[5]

The reasoning followed at the upper levels of government mirrored the sentiments expressed by ARPA's Koslov. This fact is verified by a second memo written on May 13, 1965, by B. W. Augenstein of the Office of the Director of Defense Research and Engineering to two officials at ARPA.[6] The purpose of Augenstein's memo was to urge ARPA to take on the primate project. Urging was needed because there was apparently "some internal resistance . . . to the suggestion that ARPA proceed with these experiments." Augenstein attributed this resistance to "a feeling that at one time it [the study of low-level effects] certainly attracted a number of crackpots." But while admitting that "there is some past unsavory history of experiments of this kind [the proposed primate study] in this country, which has made a number of people rather leery of further experiments in this field," Augenstein was sufficiently moved by the urgency of the embassy problem itself to recommend support. Four key points shaped his thinking.

> There is definite USIB and CIA interest in this proposition, and I believe that a USIB recommendation that such research be carried on can, or will, be generated.
>
> The existing U.S. experience in this particular energy range does not seem to be very satisfactory in quality of research.
>
> The pragmatic fact exists that the Soviet Union is irradiating our embassy in Moscow with radiation which exceeds by a factor of 100 their own safety standards, and which would give us a lever for protest if we wished.
>
> Unless, and until, other explanations are found for the purposes of the embassy radiation, this should not be left an unexplored possibility.

The seriousness with which the intelligence community took the Moscow signal created a unique situation within the policy establishment. It created a new and tightly guarded second orthodox opinion on the issue of possible low-level, athermal, central nervous system effects. Well after the Tri-Service era, majority sentiment definitely agreed with the thermal solution. This was the point of view favored by those who awarded grants, reviewed articles, and took part in scientific meetings. But the Moscow embassy problem forced a small but influential group of policymakers to break with this orthodox view and accept the possibility, if not the probability, of low-level, athermal, central nervous system effects. This split resulted in two prevailing points of view: the professional thermal point of view developed in response to the standards question and the intelligence-oriented athermal point of view stemming from the national security crisis of the Moscow signal.

The existence of two orthodoxies did not become a problem immediately. Throughout the 1960s few researchers knew anything about the Moscow embassy problem or the Soviet literature and therefore were not aware that their thinking was being challenged. By contrast those who dealt with the Moscow embassy crisis were concerned mostly with possibilities and did not challenge the thermal point of view openly, but the situation provoked by the signal did raise questions. Augenstein, for example, noted that U.S. policy had resulted "in the setting of standards of safety which are approximately 1,000 times lower [less stringent] in this country than in the Soviet Union, with our standards being set primarily by thermal damage thresholds."[7] Such looseness, based on a single assumed mechanism, may have appeared satisfactory to those who were supposedly protecting workers and the general public by setting standards, but it was not satisfactory for those charged with protecting national security. Accordingly under the cover of national security, the planned two-pronged investigation into the signal problem got underway during the summer of 1965, beginning with surveys of the health of embassy employees.

Moscow Viral Study

Plans to give special medical tests to embassy employees were made as early as March 1965. The TSCC meeting of April 15 gave these plans a boost by requiring all USIB member agencies to cooperate with the study and review the medical records of their employees in Moscow. By May 1965 letters had been issued to the doctors who would conduct the tests, and embassy personnel were given instructions for receiving medical check-ups before and after their tour of duty.[8]

Although no official justification has ever been given, U.S. officials who dealt with the signal problem assumed that it had to be kept secret. This meant that the planned health surveys had to be undertaken without disclosing their real purpose, even to the embassy employees who were the potential target of the radiation and of the health surveys. From the State Department's perspective, inventing a cover story proved to be simpler than might have been expected.

By the mid-1960s exposure to radiation had been potentially linked to both eye and blood disorders. One of the few personnel surveys undertaken during the Tri-Service era had turned up correlations between RF exposure and lens defects.[9] A paper read at the American Institute of Biological Sciences in Boulder, Colorado, in August 1964 reported finding altered blood serum counts in guinea pigs exposed to as little as 10mW/cm^2 of RF radiation, an effect said perhaps to have been due to "a microwave effect that was not heat."[10] Blood test and eye exams thus became logical measures for use in health surveys.

In addition to their potential scientific importance, blood tests and eye examinations had the advantage of being convenient and unobtrusive. Both are normally given during routine physicals. Therefore it was simple for the State Department to institute a medical surveillance program once the decision was made to take the signal seriously. Consulting physicians in Moscow and the United States were told that there was concern in the State Department over possible viral contaminants in the Moscow environment, a concern used as an excuse to institute the Moscow Viral Study.[11] Under the study embassy personnel were asked to submit to routine physical exams before going to and on returning from tours of duty in Moscow. During these exams, blood samples were taken and eye examinations administered, for the undisclosed purpose of checking for adverse effects stemming from exposure to the Moscow signal.

In the course of putting this plan into operation, several minor complications arose. The post-tour-of-duty checks had to be arranged on a tight Monday through Thursday schedule to coincide with the work program of the Washington-based doctors conducting them, but scheduling precise returns from Moscow for entire families was sometimes difficult. Later when it was decided to collect additional blood samples in Moscow, the routine trials of life there made it difficult to meet deadlines. In April 1967 the doctor in charge of collecting the samples in Moscow apologized to his superiors in Washington for delays. He was trying to assemble the basic equipment needed to collect and prepare the samples for shipment. His search extended from Weisbaden and Helsinki to the offices of his colleagues in the British embassy in Moscow. Two months later he again apologized for delays, this time because the workmen installing x-ray equipment were "digging holes in walls, stringing wires, etc., in other words, taking over my office."[12]

Shortly after the blood tests and eye exams were begun, the health survey project was broadened to include a search of medical records, as had been proposed earlier by the TSCC. By October 1965 a list of 226 persons who had served in Moscow since 1953 was drawn up. Of this number, medical records for only 139 could be located. The rest were assumed to have served in military capacities or to have left the State Department. The 139 files were then supplemented by the medical records of their dependents, giving a total Moscow sample of 407 individuals. Once assembled the records were "reviewed in detail for medical conditions, considered potentially significant, which presented [themselves] during the tour of duty, or on subsequent physical examinations."[13]

The results of the records survey did little to clarify the hazards question. Although it did turn up effects, none could be related to microwave exposure. The State Department's doctors could find "no recognizable patterns of disease or evidence of recurring problems." The lack of recognizable patterns was not significant, a fact fully realized at the time. The population being studied was too small to yield statistically significant results. No controls had been singled out for comparisons. The medical records were filled out by many doctors and hence not internally consistent. Some of the data being used were supplied by persons being surveyed, such as by parents, who simply noted that their children were "in good health."[14] In other words the medical records search provided little more than a coarse sieve that could catch occasional boulders, not stones and pebbles. The only conclusion that reasonably could be drawn once it was completed was that the signal was not causing any immediate catastrophic health problems.

This conclusion was not supported by the initial reports from the blood sample study. For reasons that have never been explained, the State Department added a test for genetic abnormality to this study in late 1965.[15] When this test apparently detected unusual numbers of abnormalities, interest in the project increased. As the contracts for the genetic study were extended through 1967, 1968, and into 1969, the urgency with which State Department officials acted grew in intensity. Top priority was given to ensuring that samples were collected on time in Moscow so that they could be delivered by a special courier pouch in the laboratory in which the tests were run in Washington.[16] Then, in mid-1969, support for the genetic tests, which were carried out by Dr. Cecil Jacobson at George Washington University, was abruptly terminated. On June 18, 1969, Jacobson was sent a terse letter by the deputy medical director of the State Department, Carl Nydell, informing him that the department would not renew the contract for the study and requesting a complete report. Jacobson's five-page summary and defense of his work arrived at the State Department on August 4, 1969, thus ending the most controlled effort to survey one aspect of the health of the Moscow embassy employees.[17]

The abrupt termination of the genetic screening program in 1969 has raised questions about the State Department's motivations. Although Jacobson has never claimed publicly that correlations were discovered—he could not make such claims since he did not have access to exposure data—others have made connections and concluded that the State Department's termination of the contract is tantamount to a deliberate suppression of damaging evidence. *New Yorker* magazine writer Paul Brodeur has characterized a report on the project by a State Department physician, Herb Pollack, as "the 'smoking gun' of the State Department's genetic coverup."[18] To determine whether such charges can be substantiated, it is necessary to discuss in more detail the methods of the genetic study and the reasons for its ultimate termination.

Once the decision was made to conduct genetic tests, State Department officials followed a familiar course of action. Since the military researchers who began the blood sample study did not have experience with the procedures needed to test for genetic damage, an outside contractor had to be called in. For State Department officials George Washington University was a familiar place to turn. They regarded the university's resident expert on such matters, Dr. Jacobson, as someone who was "handy to us and well known" in his field.[19] Jacobson proposed to use a technique that he and others in cytogenetics were then developing to assess the effect of possible mutagenic agents, such as the Moscow signal, on humans. The hypothesis that the State Department and Jacobson assumed was that if the signal were a mutagenic agent, its effects could be seen in the genetic material (the chromosomes) of the affected individuals.

The procedure Jacobson used to check for genetic damage was simple in principle and has since become widely used. The survey process began by growing the blood collected from Moscow employees on a special culture medium. The cultured blood was then smeared on slides, magnified, and photographed. A technician next examined the photographs and rated the genetic condition of the blood on a scale of 1 to 5, 5 designating extreme damage and 1 for no damage, or normal. The points in between represented various stages of chromosome damage development (table 1).[20]

Who was sampled and when was determined by the State Department. Jacobson recommended screening before, during, and after duty in Moscow so serial results could be accumulated. In this way an individual's normal condition could be established. Whether sampling was actually carried out in this way is not known since the State Department has not released the information needed to correlate the results Jacobson obtained over the next three years with RF exposure. (This information could be released in a general form without compromising medical confidentiality.)

Since Jacobson did not have data on exposure, he could not make determinations about the potential danger of the Moscow

Table 1
Scale and Interpretation Used for Rating Moscow Viral Study

Scale	Mutagenic Level	Clinical Significance	Past Reports
5	Extreme	Definite	Aneusome
4	Severe	Questionable	High risk
3	Moderate	Suspect	High suspect
2	Mild	Borderline	Questionable
1	Control	None	Normal

signal; however, he could and did evaluate the damage he saw on the basis of other experiences with such tests and drew tentative conclusions about clinical significance—that is, about the effect any chromosomal damage he saw was likely to have on the health of the person whose blood was being examined. His evaluations led him to believe that he was looking at blood samples that evidenced clinically significant chromosomal defects. When he reported these results to the State Department, he recommended that specific actions be taken:

1. Patients who repeat at level 3 or higher should not reproduce until six months after their somatic levels return to 2 or 1.

2. Patients at level 4 should be withdrawn from mutagenic exposure and monitored each month until less than 3 is obtained on two consecutive samples.

3. Patients at level 5 require close clinical supervision and must be withdrawn from the mutagenic source.[21]

The seriousness of recommendations undoubtedly prompted the State Department to continue supporting Jacobson's work through 1967 and into 1968, but when in late 1968 Jacobson asked for an additional eighteen months of support, the State Department decided to seek reviews of his work.

The need to assess the validity of Jacobson's work was genuine. His progress reports regularly placed about half of the samples being studied at any one time in the 3 through 5 (moderate to high-risk) categories. The November 1968 report, for example, placed twelve of twenty-seven samples in this range. By February 1969 seven more samples had been studied, and the number in the moderate to high-risk category had risen to eighteen in thirty-four. Although these figures did not necessarily reflect actual health problems—Jacobson assumed "the most conservative attitude toward patient protection"—if even half or a quarter of the suspected ill effects did manifest themselves, there could have been a serious health problem.[22] Some form of review was indeed logical to pursue; however, the type of review the State Department undertook is more difficult to explain.

By far the simplest way to test Jacobson's results would have been to decode the samples and determine whether any correlations existed. If the moderate to high-risk results were found distributed among controlled as well as exposed populations, it could well have been suspected that Jacobson's experimental procedure was flawed. Or the absence of correlations could have tentatively been interpreted as indicating that the Moscow signal was not producing genetic effects. But if correlations were found between exposure to the signal and moderate to high-risk ratings, then the quality of Jacobson's work would have been irrelevant. Since he did not know which samples came from exposed persons, he could not have biased the results. In addition the statistical likelihood of finding correlations where none existed would have been slight. How slight could have been determined by subjecting the reported data and decoding information to a few basic statistical analyses.

There are indications that the State Department did attempt to make correlations. A State Department copy of Jacobson's first report covering samples 1 through 30 has penciled in by two of twelve moderate to high-risk reports the word *control,*[23] perhaps indicating that Jacobson was finding significant correlations and that more than chance was involved, assuming that the other ten samples came from the exposed populations. (If Jacobson's techniques were flawed, the control-exposed ratio should have approached fifty–fifty, assuming the control-exposed populations were equal.) Correlations also seemed to be on the mind of the State Department's medical consultant, Herb Pollack, when he reported to colleagues in April 1969 that of "one-hundred-forty blood samples . . . examined over a four-year period . . . four reportedly showed serious chromosomal abnormalities."[24] How the figure of 4 in 140 was derived or its significance is not certain. It could represent the total moderate to high-risk samples that fell in the exposed population or simply Pollack's own interpretation of *significant.* But there is no indication that correlations, or the lack of them, comprised the State Department's main evidence in judging the validity of Jacobson's work.

Instead, when the need for review became apparent, the State Department simply bundled up some of the slides that Jacobson had made and shipped them off to other researchers for independent assessments. This procedure produced equivocal results, which the State Department should have expected from the initial feedback it received. The first reviewer, Dr. David Hungerford of the Institute for Cancer Research in Philadelphia, complained about the quality of the slides he was sent, so much so that he said he would never agree to undertake such a project again. He rated "the best slides in this series as no better than average." His complaints about quality did not necessarily invalidate the results, however. If anything, Hungerford felt that "the data are almost certainly underestimates of the time frequency of the various effects on chromosomes." Moreover, while not impressed with Jacobson's 1 through 5 rating system, he did agree with the "recommendations for withdrawal from 'mutagenic exposure,' " although he would not have proposed "any specific reproductive counseling made on the basis of observations on a single type of somatic cell." In sum, although there were problems, Hungerford's review did not call for an end to the tests. He did recommend that their quality be improved under the existing contract or else be transferred to another laboratory.[25]

A second review submitted by a cytogeneticist working at the National Institutes of Health, Joe Hin Tjio, raises even more uncertainty. In a report submitted on January 6, 1969, Tjio rated four slides that Jacobson had put into the marginal category (2.5–3.5) as healthy. Another slide that Jacobson had also put in the marginal category (3.0), Tjio rated as unhealthy, adding in a parenthetical note: "Is it a leukemia patient?" Thus, one of Tjio's evaluations was more severe than Jacobson's, others less so. Nearly ten years later Tjio supposedly summarized these results in a brief note to Pollack and concluded, "In 1968–69, we studied the karyotypes of cultures of several blood samples from the State Department. We did not find any abnormalities. Subsequently, we reviewed the slides prepared by Dr. Jacobson which we found inadequate for karyotypic analysis. We disagreed with Jacobson's conclusions and prognostic comments."[26] How Tjio rationalized the possible discovery of a leukemia patient with no abnormalities is not indicated in his summary.

A third reviewer, Dr. Victor McKusick of Johns Hopkins University's Medical Genetic Division, agreed in February 1969 to look at twenty slides and evaluate them. In a conversation prior to the agreement, McKusick made it clear to State Department officials that "the percentage of breaks in chromosomes which would be considered significant depends on the lab doing the work and may change from time to time even in one laboratory due to culture techniques." A few weeks later he withdrew from the review process, noting that "it simply is not possible in this very tricky field to go about it [reviewing the slides] in this way."[27]

When State Department officials summarized these reviews, they focused on the negative comments about experimental technique. In a memo intended for the in-house review process, three points were listed as the "Consensus of Opinions": (1) the slides were of poor quality, and the number of observations made were not sufficient to draw (2) scientific or (3) clinical conclusions.[28] This one-sided negative assessment was used to justify an action the State Department had contemplated taking as early as March 17, 1969: "scrap it [the genetic project] entirely."[29]

George Washington University's dean of the school of medicine, Dr. John Parks, questioned the decision to terminate Jacobson's work, especially under the apparent cloud of negative reviews. The State Department thus agreed to seek one more outside opinion, from Dr. Kurt Hirschhorn of the Mount Sinai School of Medicine in New York City. Hirschhorn was sent thirteen slides to study; they were not the same ones reviewed by either Hungerford or Tjio. (There was no indication in the documents made public that two reviewers ever reviewed the same slides so that their opinions could be checked.)[30] Most of the slides Hirschhorn was sent fell into Jacobson's marginal range, although it is difficult to be sure of this fact since all of the ratings have not been made public. Hirschhorn reportedly found fewer chromosome breaks than had been reported by Jacobson and included most of the slides in the normal range for his laboratory.[31] This assessment, which was only one opinion, reinforced the State Department's belief that they could not rely on Jacobson's reports, and his support was terminated.

The State Department's use of reports critical of Jacobson's work to terminate the Moscow viral study is difficult to justify. The responses received were neither consistent nor replicated, thus leaving the question of validity unanswered. It may have been the case, as reported at one Pandora meeting, that "the evidence [of chromosomal abnormalities] was based upon weighted data, which may not be acceptable to all experts in the field."[32] But the State Department had also been told that it could not test the validity of Jacobson's work by simply asking others to reevaluate his slides. Had there been no other way to evaluate the study, the approach taken might have been used as a last resort, but there were other ways to proceed.

Jacobson and the State Department entered into a typical "if this, then that" scientific experiment when they began the genetic phase of the study. The basic premise of the experiment was that if the Moscow signal were a mutagenic agent (could damage chromosomes and so perhaps cause birth defects or cancer), its effects could be measured as a statistical difference between health-related conditions in control versus exposed populations. Having begun the work with this premise, the significance of the experiment could have been tested only by finding out whether statistical differences existed. Once differences were found, or not found, the next task in assessing the importance of the signal would have been to determine the significance of the statistical differences. It is at this point that the quality of the slides and discussions of clinical implications would have become important. But until the basic issue of statistical differences was settled, seeking the reviews the State Department sought served no legitimate scientific purpose. Results cannot be tested until they are reported, and in this case reporting required completion of the experimental program to find out whether anything unique was appearing in populations exposed to the Moscow signal.

The fact that the complete results of Jacobson's work have never been made public means that it is impossible to conclude whether Pollack's comments about 4 cases of damage in 140 samples is a "smoking gun." Also the widely reported statements by Dr. Thomas Gresinger, a resident at George Washington University at the time Jacobson was conducting his tests, about "funny results" and "lots of chromosome breaks" prove nothing.[33] Nothing can be proved until more information, not only on Jacobson's studies but on the entire health survey project, is made public.

At the same time that Jacobson was testing the Moscow blood samples for genetic damage, physicians at Walter Reed Army Hospital were testing similar samples for evidence of other effects. The results of these tests have not been made public.[34] Screening for behavioral effects was done on embassy employees in 1968, but again the results have not been made public.[35] There are, in brief, many unanswered questions about the early health studies, their purpose, and their findings. Until these questions are answered, different observers will draw different conclusions from the information that is available. If some of these observers continue to remain skeptical about State Department motivations, it must be admitted that although they cannot prove their case, their case also has not been disproved.

Pandora

The second investigative effort into the signal problem, following USIB and White House recommendations, focused initially on one major project: subjecting monkeys to a simulated Moscow signal. Work on this project began in October 1965 under the code name Pandora.

In describing the work done under the Pandora code name once it became known to the public in 1976, government and military officials denied that it was undertaken in any weapons context. In 1976 ARPA's director, George Heilmeier, responded to a mailgram addressed to President Ford by assuring the sender that "the Pandora experiments were never directed at the use of microwaves as a surveillance tool nor in a weapon concept."[36] A year later Heilmeier expanded on this point in a written response to a congressional inquiry: "This agency is not aware of any research projects, classified or unclassified, conducted under the auspices of the Defense Department, now on-going or in the past, which would have probed possibilities of utilizing microwave radiation as a form of what is popularly known as 'mind control.' "[37] Neither of these statements is historically accurate. Weaponry and mind control were clearly on the minds of some ARPA officials when Pandora began.

Direct central nervous system effects were taken seriously enough to prompt convening the 1963 UCLA conference. Although the tone of this conference generally downplayed the possibility of such effects, discussions on this topic did continue. In August 1965 a biophysicist at Northrop Space Laboratories in California prepared a technical memorandum, "Biological Entrainment of the Human Brain by Low-Frequency Radiation," in which he speculated that if it were true that the body's biological clock were related to electrical impulses in the brain known as alpha rhythms, it might be possible to alter that clock by subjecting it to specific electromagnetic fields.[38] A copy of this paper made its way into the files of the acting director of the Advanced Sensors program at ARPA, Richard Cesaro, who was responsible for initiating project Pandora.

The impact of such scientific speculations on Cesaro's thinking is evident in his initial description of Pandora. He justified the project by noting that "little or no work has been done in investigation of the subtle behavioral changes which may be evolved by a low-level electromagnetic field." Researchers had established that direct electric stimulus of the brain could alter behavior. The question raised by RF radiation was whether the electromagnetic energy created by its fields could have a similar effect at very low levels. This is what project Pandora initially set out to discover: whether a carefully constructed microwave signal could direct the mind.[39]

The team assembled to answer the behavioral effects question included civilian as well as military scientists. The army's Walter Reed Institute of Research provided the laboratory, the chimps, and some of the biological expertise. The air force assembled the RF equipment and oversaw its running. Advisers from the Johns Hopkins Applied Physics Laboratory tested the RF equipment to make sure that it produced the desired signal at the proper intensities. Finally additional consultants were contacted to advise on specific problems, among them Maitland Baldwin of the National Institutes of Health, Ross Adey, a sponsor of the 1963 UCLA conference and well-known brain researcher; Milton Zaret, an ophthalmologist who had done personnel surveys looking for eye problems during the Tri-Service program; Jo Johnson, the CIA's person in charge of monitoring the Moscow signal; and Herb Pollack, then with the Institute of Defense Analysis and who played a key role in the Moscow Viral Study.[40]

It took about a year to assemble this research team and set up the Walter Reed test facility. By October 1966 the last of the RF equipment was in place, and the Applied Physics Laboratory advisers had prepared an instruction manual for its operation. One month later the team issued its final report, certifying that the equipment was capable of generating the desired signal at specified power levels.[41] Reports from Moscow were still putting the intensity of the signal in the 0.5–1.0mW/cm^2 range and in the low gigahertz frequency band. The Walter Reed test facility was thus designed to study the effects of a 2.2–4.0GHz modulated signal at power levels up to about 5mW/cm^2, the maximum power of the equipment being used.

As soon as the RF equipment was cleared for use, the first chimp was brought into the exposure chamber and testing begun. The effect of the stimulated signal on the chimps was measured by training them to perform specified tasks and then observing how well they performed those tasks during prolonged exposure. After only four weeks of testing, startling results began to emerge.[42] During the first week of ten hours of daily exposure (roughly the amount a worker in Moscow might receive), all seemed normal. The chimp continued to perform normally into the second week. But on the twelfth day "a definite slowdown was recorded in the monkey's ability to time his work functions. On the thirteenth day of radiation, the monkey further degraded and finally stopped working." For all intents and purposes the test animal appeared to be in a deep sleep. This condition persisted for two more days, after which the signal was turned off.

Once removed from the signal the chimp returned to its normal work routine. Over the next five days, nothing unusual was observed. Then the power was turned on again. This time it took only eight days for the symptoms of slowdown to appear, and in ten days the animal had returned to a deep sleep condition. The chimp was again allowed to remain in this state for three days before the power was turned off. Ending the exposure this time did not end the slowdown, however. Project director Cesaro waited two days and then, on December 15, 1966, prepared a report. As he was writing, recordings indicated that "the monkey had not returned to normal."[43]

These admittedly sketchy results had an immediate impact. A new code name—Bizarre—was assigned to the project and security tightened even more. Cesaro raised three specific issues that had to be considered in more detail. First, "the potential of exerting a degree of control on human behavior by low-level microwave radiation" seemed to exist. Second, the fact that the Soviets had studied such effects in some detail whereas "only isolated investigations have heretofore been carried on in the U.S." could not be ignored. And third, since it now seemed possible that effects might be found at exposure levels below 10mW/cm^2, there was an obvious need to overhaul "U.S. microwave radiation safety standards . . . to take account of the non-thermal damage potential."[44] This last conclusion was reached one day short of a month after C95.1-1966 was formally adopted and the broad standard-setting community, not privy to Pandora results, assumed that the bioeffects problem had been solved.

Given the early test results, there was no doubt that Pandora had to be continued and intensified. The obvious next step was to run trials on a second chimp. If similar results emerged, further tests would have to be designed to determine how the

RF radiation was affecting behavior. Cesaro felt that "there is no question that penetration of the central nervous system has been achieved, either directly or indirectly," but no one understood how this was happening, if in fact it were.[45]

As more data were assembled over the next year, the intense concern produced by the first month of experimentation lessened considerably. Continued monitoring of the signal in Moscow showed that the initial power estimates of 0.5–1mW/cm^2 were much too high. On September 13, 1967, the CIA sent a secret memo to ARPA's director of R&D putting power levels in the low-microwatt range. This changed the situation entirely, since it now seemed that "the Soviets have never transmitted the 'signal' at a level above their exposure standards."[46] When tests were run on chimps at the lower levels (0.008, 0.05, and 0.01mW/cm^2), no degradation of the normal work patterns was observed.

The disappearance of the mind control explanation for the signal turned project planners to more subtle lines of testing. If the effects of the signal were not immediate, perhaps they were latent or long term. U.S. researchers had not studied long-term, low-level effects; the Soviets had. There still remained the issue of the Soviet work and its findings. In addition the absence of immediate central nervous system effects did not rule them out altogether. Cesaro still felt that central nervous system effects could be important and urged their study "for potential weapon application."[47] Whether for weapon application or not, central nervous system studies did continue under Pandora sponsorship, along with efforts to discover whether long-term effects were possible and whether there was any useful information in the Soviet literature. With the addition of subsequent projects, Pandora became a multimillion dollar research program with a number of projects.

1. Regular reports were received from Allied Research Associates in Massachusetts on a computer-based "review, analysis, and classification: of the USSR bioeffects literature." The primary researcher was Janet Healer, who had been working for Allied and ARPA's Sam Koslov for about three years developing projects related to the exploitation of sense-perception literature for weapons development.

2. Under the code name Project Big Boy, Herb Pollack, Joseph Kubis (a psychologist at Fordham), and others conducted a battery of tests on sailors on the aircraft carrier

Saratoga to determine whether personnel who regularly worked above deck, in presumed high RF exposure areas, tested differently from personnel who worked below deck. These tests were conducted during early 1969 and summarized in a final report by Kubis in May 1969.

3. Extensive discussions took place and plans were made to extend the primate tests to humans.

4. The chimp studies were broadened to include tissue analysis and readings on electroencephalogram and heart rate. The purpose of the new studies was to attempt to understand how RF radiation might be affecting the central nervous system.

Contracts on related research projects, many of which were being supported with military money channeled through normal funding paths, were also regularly reviewed for pertinent information. Some of the projects that fell under this category were Jacobson's genetic studies being done for the State Department, epidemiological surveys of possible birth defects in the children of Korean War veterans that were done under the supervision of Dr. Abraham Lillienfeld at Johns Hopkins, Adey's brain research at UCLA's Brain Institute, Milton Zaret's attempts to duplicate heart effects reported by Soviet researchers, and a study of athermal genetic effects in plants being done at a private research foundation in Connecticut, the New England Institute for Medical Research, under Dr. John Heller.

The progress made during the expanded phase of Pandora experimentation was reviewed by a new civilian Science Advisory Committee formed in December 1968.[48] A year and one half of intensive review undertaken by this committee and the multimillion dollar research program that went with it did little to clarify the signal problem. At every point equivocal results emerged. The literature search conducted under Healer's guidance turned up many reports of low-level, athermal effects, but few, if any, of these reports were specific enough to allow them to be accepted with confidence. But they could not be dismissed, especially since the signal not only was continuing but also being changed by the Soviets from time to time. The Soviet literature contained no obvious keys to open or lock the door on the bioeffects and signal problems.[49]

Project Big Boy turned up no significant differences between the responses of above- and below-deck personnel to an extensive battery of psychological tests. Had the exposure situations of the above- and below-deck personnel been different, as initially anticipated, this would have given exposure to shipboard radar a partial clean bill of health, but when careful measurements were made, no significant exposure differences could be verified. Therefore the *Saratoga* study added no new information,[50] a fact frequently overlooked in later summaries.[51]

Efforts to extend the chimp tests to a small group of humans began at the first Science Advisory Committee meeting in December 1968. Plans called for subjecting a small group of healthy young males to the simulated Moscow signal for at least ninety days, unless significant effects appeared before then. During this time the test subjects were to be monitored for everything from blood changes, electroencephalogram, and blood pressure to modifications of behavior and chromosomal aberrations. The need for genetic testing was undoubedly influenced by the controversial results emerging from Jacobson's work and the fact that similar tests run on "the peripheral blood of one monkey, previously irradiated with the 'Moscow signal' at a field strength of 4.6mW/cm^2, indicate marked aberrations in 40 percent of the cells."[52]

No work was done on the human study in early 1969. In April 1969 the Pandora panel met and specifically recommended that the Walter Reed facility develop a human program. Fort Dietrich was targeted as a likey place for volunteers. Suggestions on control procedures were advanced, and one precautionary note was added: "shielding of testicles is recommended."[53] Clearly no one was prepared to take any chances with genetic effects.

During the next four Pandora meetings (May 12, June 18, July 16, and August 12–13), discussion of the human study continued. Plans were made to construct a two-chambered exposure facility at Walter Reed that could house both control and power-on testing. The turn-on date for the experiments was provisionally set for October 1969, if all of the equipment could be assembled. But then a hitch in these plans developed. During this entire period the Pandora committee had also been discussing the significance of the findings in the chimp studies. In the course of these discussions, considerable disagreement emerged, with one member of the Pandora team arguing that significant effects had been found and another arguing that reanalysis showed that no significant effects had emerged. Given these differences, the Science Advisory Committee had by August 1969 turned all of its attention to the evaluation of the primate study, leaving the human studies in limbo.

The need for a reanalysis of the data from the primate study became apparent in August 1969 when a new director assigned to the project, Major James McIlwain, challenged the conclusions reported previously. McIlwain had worked on the chimp project under its former and only other director, Major Jo Sharp, and stepped into the position of director when Sharp left the military in early 1969. On assuming command, McIlwain reanalyzed the data that had been collected and challenged the conclusion that meaningful correlations between RF exposure and effects had been found, a fact he reported to the Science Advisory Committee in August.

McIlwain's report radically altered the course of the Pandora program. Committee members attended the August meeting with the expectation of brainstorming the RF bioeffects question in an effort to get some immediate answers to the Moscow embassy problem. By mid-1969 Pandora was moving into its fifth year, and still no firm conclusions had been reached. Higher up, patience was wearing thin. Earlier in the year Cesaro essentially told the committee members that they had one more year "to make major advances in solving the [embassy] problem."[54] To meet this call for decisive action, two days instead of the customary one had been set aside for the August meeting to allow time to get Pandora on a track that would lead to some concrete answers. Instead McIlwain's report led to even greater uncertainty. Until this uncertainty was resolved, "the Committee felt that the first order of continued effort . . . should be based upon a resolution of the apparent differences in the interpretation of the WRAIR primate program."[55]

The committee turned to one of its members, Joseph Kubis, for an evaluation of the primate study. Kubis was familiar with behavioral studies, knew how to employ statistics in testing, and had proved himself to be an even-handed investigator during Project Big Boy. In late October 1969 he was asked by Pandora's chairman, Lysle Peterson from the University of Pennsylvania, to develop a program for evaluating the primate studies. Over the next month Kubis met with all of the principals involved, studied eight notebooks containing the raw and analyzed data, and drew this information together in a preliminary report dated December 4, 1969.[56] This report demonstrated conclusively and with unusual sensitivity to the complexity of scientific research that the primate studies had not and probably would not in the near future provide a conclusive solution to the embassy problem.

The interpretive problem that surrounded the chimp studies had its origins in two subproblems: the complexity of the tests themselves and contrasting understandings of their main objective. Jo Sharp, who had helped design the project, saw his basic mission as exploratory in nature. The main question he asked himself as he looked at the data was whether he was seeing anything that might be a behavioral effect. His objective was not to prove that a specific effect existed but to discover whether any of several effects might be possible. From tentative discovery, research could then progress to proof and an understanding of mechanisms.

McIlwain, who came to the project only after it had been going for a while, had different priorities. When he looked at the data that had been collected over the years, the first question he asked was whether they proved anything. By proof he meant drawing from the data irrefutable statistical correlations. The complexity of the experiments, however, made it almost impossible to produce such correlations.

The primate studies were not as simple as the early reports issued by Cesaro or as popular accounts suggest. Before the chimps were exposed, they were trained to perform a complex work cycle over the course of an eight- to ten-hour day with the goal of gaining food pellets. In order to get pellets, the chimp had to push a food level in accordance with a preestablished program learned during the training period. One type of buzzing noise indicated to the chimp that it was in a time-out period; if it waited ten minutes and pressed the lever it would get food. If it pushed the lever after only nine minutes, the whole cycle would start over, requiring an additional ten minutes.

After it earned a pellet in this way, a red light came on indicating that the time-out period was over and the progressive ratio period begun. During this second period, the chimp had to push the bar 40 times to earn one pellet, 80 times to earn a second, 160 times for a third, and so on until 640 presses yielded a fifth pellet. When this task was completed, a second time-out phase was entered, and so on until the animal earned the number of pellets that had been programmed for a normal day's consumption—70 to 100 pellets, depending on the animal's weight.

Once the chimp had learned to perform these tasks in the training chamber, or ice box as it was called, it was placed in the RF exposure chamber, the power turned on, and detailed records automatically kept on performance. Ideally changes in performance at this stage should have indicated effects stemming from RF exposure since this was the only parameter being changed. But RF exposure was not the only parameter that had been changed. The ice box was not the same as the exposure chamber; preconditioning had been done during the day; RF testing was frequently done at night (to avoid security problems and RF interference); new chairs and new food levers replaced old ones; cages were a different color; solitary confinement replaced group conditions; and so on. Then there were other problems that developed: feed tubes jammed or feeders became empty, ink tubes on automatic recorders dried up, the power went off. Any one of these changes could have affected behavior. The problem researchers faced in interpreting the results was finding out whether through all of the resulting morass of data, any meaningful patterns emerged.

From a discovery point of view Sharp believed that there were correlations worth further study. McIlwain felt that the prior tests proved little or nothing at all; he personally found no statistical correlations. Kubis agreed that much of what Sharp thought was interesting had not been proved, disagreed with the logic of some of McIlwain's statistical analyses, and suggested ways to get around existing problems by reanalysis or additional, carefully controlled testing.

Kubis' report came to the full Science Advisory Committee in January 1970. This final piece of evidence, like every other piece of evidence, was inconclusive. There was the obvious fact that four years of work with a simulated Moscow signal had not produced any noticeable effects. Most of this work had been conducted at 4.6mW/cm^2, a level now known to be nearly 1000 times higher than the signal itself. Common sense seemed to dictate that bioeffects were not the primary objective of the Moscow signal. But many research avenues had been left unexplored. The experimental design errors in the primate study and Big Boy could have been eliminated. The primate studies had used only a portion of the Moscow signal, perhaps the wrong portion. Human studies still could be done, although they would raise the same questions as the primate studies and bring in ethical questions as well. And to add to the uncertainties, Pandora's Science Advisory Committee had no idea how important to national security this matter actually was. They had not been made privy to the deliberations going on in the intelligence community and knew nothing of work that might be looking into other possible reasons for the signal itself. Thus the Science Advisory Committee left the future of Pandora unresolved at their last meeting in January 1970. If it was worth answering all questions, then more work could be undertaken. Worth in this case had to be determined by the national security community.

When the national security community met in early 1970 to consider the recommendations of the Science Advisory Committee, they concluded that Pandora should be terminated. The reasons for this decision are not given in any of the documentation made public. Brodeur and others have suggested that Pandora was terminated because it had discovered serious biological effects that needed to be covered up.[57] However, given the ambiguity of all Pandora data, it is unlikely that Pandora's sponsors, who characteristically were not troubled by unsubstantiated reports of hazards, felt that there were health problems that needed to be covered up when they made the decision to terminate the program. Ten years had passed since the end of the Tri-Service program and still "no one was dying." The primate study had produced at best only subtle changes. And as more was learned about the Moscow signal, it seemed that it was intended for purposes other than biological effects.[58] As a result, on July 1, 1970, Pandora was officially transferred from ARPA to the Walter Reed Institute of Research and given $200,000 for fiscal year 1971 "as final funding" for the project.[59] By early 1971 all Pandora work was absorbed back into the normal funding patterns of RF bioeffects research.

Conflicting Motivations

The termination of Pandora did not meet with universal approval, either at the time or subsequently when it became known to the general public. Some members of the scientific community were troubled by the unexplainable effects they had discovered. There were possible connections between modulation and central nervous system effects that needed to be explored. The epidemiological data that had been collected turned up possible correlations between RF exposure and effects. Until these loose ends were tied, scientists would have difficulty discounting hazards completely. This was particularly the case for one member of the Science Advisory Committee, Milton Zaret, who had a major project terminated during the closing years of Pandora. Zaret's dissatisfaction with the termination of Pandora was sufficient to turn him into a major critic of microwave policy.

The termination of Pandora also left unanswered questions about standards. Controversial findings at 4.6mW/cm^2 could be discounted in evaluating the significance of the Moscow signal since the Moscow signal was 1000 times less powerful. But 4.6mW/cm^2 was less than another important figure, the 10mW/cm^2 ANSI and military standards. Obviously all of the Pandora testing had been done at levels below the U.S. standards, a fact that did not go unnoticed. In September 1967 the CIA's Jo Johnston observed in a memo summarizing the early results of Pandora, "For the record it should be noted that all the positive findings of Project Bizarre [Pandora] were achieved one half an order of magnitude below the accepted U.S. standard for safe exposure."[60] Pandora and the Moscow Viral Study raised questions about standards that would eventually have to be answered.

The unanswered scientific questions and the issue of standards might have prevented the termination of Pandora if Pandora were primarily a study of health effects. However, ensuring the well-being of a few people in Moscow seems not to have been the primary motivation for the secret response to the Moscow signal problem. No thought was ever given to limiting exposure, even when the evidence seemed to point to serious effects. Pandora's sponsors were interested only in applied information that would provide a simple answer to the bioeffects questions raised by the signal. They needed this information for national security purposes. They were willing to go to great lengths to aid the scientific process if it would supply answers but were not interested in science in and of itself or in maximizing health protection.

The conflict between the pure and applied scientific positions emerged again and again during project Pandora. At its April meeting, when the Science Advisory Committee reviewed the work of the outside consultants, projects that strayed too far from applied objectives were singled out for possible termination. Ross Adey, for example, was to be "informed that his own priorities and work trends are not entirely matched with those of ARPA. Although his work is related to the general field and is of considerable assistance to the Walter Reed effort, it is thought that his support might be phased out in a year or two after the Walter Reed facility is better developed."[61] A similar warning against straying too far into the seductive lair of science was relayed to the Science Advisory Committee in June 1969 by Cesaro:

> Investigative programs should be designed to take major bites at the problem to achieve definite indications of whether or not there are effects on humans of microwaves under conditions simulating, as closely as appropriate, the Moscow signal. Furthermore, he urged that the experimental programs be relevant to the Pandora problem and provide significant results, negative or positive. He reminded the Committee that, conversely, it would be inappropriate to follow paths which while they might be interesting scientifically, would not be relevant to the problem.[62]

The Department of Defense did not "support . . . basic research, unless it can be shown to have relevance to national defense."[63] As soon as the bioeffects problem seemed unrelated to national defense, Pandora was terminated and bioeffects research once again left to fend for itself.

This time, however, the scientific forces were not left without supporters. Between 1960 and 1970 much had changed. Vietnam lessened public confidence in official bureaucracies, especially the military. Pollution and environmentalism had become watchwords. And the public's elected representatives in Congress had discovered the microwave problem. When they did, the microwave debate was forced to leave the security of limited policy circles and its relatively private years and moved reluctantly out into the public forum.

II

THE PUBLIC YEARS, 1967 TO THE PRESENT

6

The Search for Government Solutions

I have often heard technical management likened to a log plunging down a turbulent mountain stream with three hundred ants as riders, each of which is convinced that he is steering.—Alvin Burner, *Biological Effects of Microwaves*, 1968

The microwave debate did not become a public issue until the late 1960s, for a number of reasons. During World War II the country had more on its mind than possible subtle effects from exposure to radar. Immediately after the war RF technology rapidly assumed a key role in the Cold War, again diverting attention away from possible hazards. By the mid-1950s electronics had come to occupy such a crucial role in strategic planning that an article in the USSR's *Red Star* quoted in U.S. newspapers could boldly suggest that future generations would look back on the postwar era as the radioelectronic age, not the atomic age.[1] In such an atmosphere benefits clearly outweighed risks, and the bioeffects issue was largely ignored by the press and public.

Well into the 1960s, when more information about biological effects became available, the press usually brought it to the public's attention in brief factual reports. The beginning of the Tri-Service program was announced under the headline, "Air Force Studies Hazard of Radar." The accompanying article informed readers that the biological effects of microwaves were unstudied and therefore had become the focus of an intensive joint military-university research program. Four years later the final results of the Tri-Service program were greeted with an air of cautious optimism. *New York Times* science reporter John Osmundsen, who attended the yearly meetings, was as intrigued by the beneficial medical applications of newly discovered low-level effects as he was by hazards. His final report left readers with the reassuring note that although the standard then in existence had not been changed as a result of Tri-Service research, there probably was nothing to worry about; "if anything, the limit appeared to be on the conservative side."[2]

But values were also changing in the mid-1960s. The Vietnam war was undermining the credibility of the government and military. The civil rights movement was stalled. Environmental problems were casting doubt on the ability of industry and government to protect the environment from pollution. Inflation and economic displacement seemed destined to turn the guns and butter strategy into a guns or butter confrontation. On every front old values were being challenged.

The new mood entered the RF field through an unexpected turn of events. In May 1967 GE reluctantly announced to the public that 90,000 of its new colored television sets leaked ionizing radiation. A design error had placed a regulator tube in a position that caused it to emit X-rays through the bottom of the set. Although the sets allegedly posed no danger to public health, they were nonetheless recalled.

The physical problems created by the misdesigned television sets were easily eliminated by redesigning the sets. The psychological trauma the episode created was more difficult to overcome. Questions immediately began to arise about the apparent lack of legislation and regulations to deal with such problems. In mid-August 1967 the House Subcommittee on Public Health and Welfare called the first of a series of hearings designed "to provide for the protection of the public health from radiation emissions from electronic products."[3] Two weeks later the Senate Commerce Committee followed suit. After decades of neglect and disorganization, a concerted effort seemed underway in late 1967 to deal with the environmental and health problems posed by all forms of electromagnetic radiation, from X-rays to microwaves and the remainder of the RF spectrum. With Congress's entry into the arena, the microwave debate became a public debate.

Legislative Actions

The GE incident had not caught the government completely unaware. A few years earlier advisers in the Johnson administration had realized a need to regulate the rapid growth of industrial and civilian uses of RF technology in communication. On December 6, 1963, Johnson's special assistant for telecommunications, Jerome B. Wiesner, wrote a letter to the IEEE's Joint Technical Advisory Committee (JTAC) asking for help:

> Many organizations, scientists, and engineers are concerned with this problem [of spectrum management]; foreseeing the development of numerous situations in which the usefulness of radio frequency resources of this nation and world may be seriously limited. . . . Prompt attention should be directed to the growing problems of "spectrum pollution" by unwanted and/or unnecessary radiations.[4]

JTAC responded by forming a Subcommittee on Electromagnetic Compatibility, which labored for five years in an effort to provide the government with comprehensive guidance on spectrum management.

Although the subcommittee had not been directed to study RF bioeffects, it did assign one of its task groups the job of surveying electromagnetic side effects, "those problems of community living, *other than* interference in communication, arising from electromagnetic radiation," a definition quickly understood to encompass RF bioeffects and potential hazards.[5] In practical terms JTAC's side effects task force had little impact on the microwave debate. Its contribution to the final report issued by JTAC's Subcommittee on Electromagnetic Compatibility. *Spectrum Engineering, the Key to Progress,* was brief and lacked a comprehensive assessment of the RF bioeffects problem. Moreover, in March 1968, when *Spectrum Engineering* was published, its general conclusions on RF bioeffects were not particularly newsworthy. By this time Congress had been at work on the radiation problem for many months and was about to attempt to bring order to the field through a series of sweeping legislative reforms.

It took less than two months for Congress to draw up legislation in response to the disclosure of the GE television set problem in May 1967. On June 13, 1967, Congressmen John Jahrman of Oklahoma and Paul Rogers of Florida introduced the Radiation Control for Health and Safety Act of 1967 (HR 10790) on the floor of the House. A week later similar bills (HR 11101 and HR 11107) were introduced by David Satterfield of Virginia, L. H. Fountain of North Carolina, and John Dingell of Michigan. The Senate followed suit on July 10 with two similarly titled bills (S. 2067 and S. 2075) sponsored by senators from Alaska, West Virginia, Idaho, Hawaii, Maine, and Rhode Island. By mid-August ten bills had been introduced responding to the presumed radiation crisis.

The urgency that underlay this activity was more than evident at the time. Congress had been remiss in the past, but it would not be so in the future. As Senator E. L. Bartlett commented when opening the Senate Commerce Hearings on August 28, 1967, "We are working in an area in which the exercise of foresight is essential. We have all been sobered, I think, by the recent near crisis involving television receivers, and we have come to realize that this is an area in which we cannot proceed on a crisis-to-crisis basis. This is an area where stop gap measures will not suffice. We must, rather, take steps now to insure that dangerous situations will not arise and that long-term and far-reaching damage will not be done."[6] To achieve the desired end of maximum public protection, witnesses were scheduled to testify before the House and Senate committees reviewing the proposed legislation. Each witness had as his or her primary objective educating Congress on some aspect of the radiation problem so that an informed decision could be made.

Although RF radiation took up only a small percentage of the time during the 1967–1968 congressional hearings on the pending bills, information about RF bioeffects did make its way into the hearing records. Congress learned about the sharp discrepancies between U.S. and USSR exposure standards, about the rise and fall of the Tri-Service program, about the paucity of legislation governing the use of RF equipment in the public sector, about the uncertain nature of health hazards (cataracts and Down's syndrome were two primary concerns), and about the lack of scientific knowledge on RF bioeffects.

The partisan points of view entered into the record by the designated experts ensured that Congress received conflicting information. A letter from Raytheon's Microwave and Power Tube Division in Lexington, Massachusetts, confidently informed Congress that there was no problem to solve. Looking back on the Tri-Service era, Raytheon's lawyer concluded that "this program, begun in the late 1950s, reached three basic conclusions: 1) The biological effects of microwave energy are thermal. 2) The effects are non-cumulative. 3) Man has a built-in alarm system coupled with his threshold of pain that protects him from thermal injury." This being the case, he advised Congress that "although there are, in our view, ample standards now available to industry, their codification in a single instrument might be helpful. However, it is not thought that extensive investigation to arrive at such standards, on which so much money has already been expended, would be needed."[7] In other words, since in Raytheon's opinion the only possible ill effect humans could experience in RF fields was temporary overheating, which could be detected by the body's natural sensing system, Congress was advised to forgo sponsoring further research and to get on with the task of establishing a uniform public protection policy based on existing information.

A different point of view was presented by the person who had been instrumental in setting the $10 mW/cm^2$ standard on which industry so confidently relied, Herman Schwan. In Schwan's opinion, "the present standard, while a significant step forward, is only a beginning in the gradual evolution of a set of standards, which cover all circumstances of practical interest." Much more had to be learned. There was not enough evidence to limit RF bioeffects solely to thermal mechanisms. Schwan could not discount genetic and cumulative effects completely. While stopping short of arguing that "there is a very real danger of genetic damage," he was willing to suggest that "there is a possible danger. We don't know about it." As a scientist he advised Congress to foster more basic research and to encourage data collection.[8]

Other experts presented different views. A specialist in eye research at Tufts University, Russell Carpenter, informed Congress by letter that he had "clearly demonstrated a cumulative harmful effect of microwave radiation on the eye [of rabbits]."[9] The military felt it had the situation well in hand, so much so that in time its "regulations will evolve to be less restrictive and more discriminating as greater knowledge and experience will permit."[10] A prominent radiologist from Johns Hopkins argued strongly for excluding microwave radiation from the pending legislation since "from the standpoint of a health hazard, ionizing radiation represents the problem area of dominant importance." In his opinion the only health hazard microwaves posed was that associated with superficial burns.[11]

Such conflicting testimony notwithstanding, by early 1968 the Democratic-controlled Congress received a clear message from the White House indicating that it was expected to do something about the radiation problem. In his 1968 State of the

Union Address, Lyndon Johnson pledged to keep Americans "safe in their homes and at work." Part of the new deal consumers were being promised was "protection against hazardous radiation from television sets and other electronic equipment."[12] On October 18, 1968, the promised protection became law when Johnson signed a slightly modified version of the original 1967 House bill, now called the Radiation Control for Health and Safety Act of 1968 (PL 90-602).[13]

PL 90-602 accomplished only one thing: it dumped the entire radiation problem, RF bioeffects included, in the lap of the secretary of Health, Education and Welfare (HEW). The secretary was told to "protect the public health and safety . . . from the dangers of electronic product radiation" but was not given much guidance for setting about this task.[14] Full authority was granted to conduct surveys, purchase and test electronic equipment, hire consultants, contract research, and cooperate with other private and government agencies—in short, to do anything that would result in public protection. Little guidance was given on the truly difficult issues. Public protection was not defined. No specific problem areas were identified.

The drive for improved public protection continued after Lyndon Johnson left office under the cloud of the Vietnam war and eroding public confidence in government at home. Richard Nixon's narrow victory in 1968 saw little change in the composition of Congress or retreat from a commitment to the program of social reform begun during the Kennedy-Johnson era. The conservative resurgence that a decade later would see government's role in regulation severely challenged had yet to manifest itself in a major way, although there were signs that a swing toward less and not more control was underway.

The continued drive for social reform and greater public protection during the remainder of the 1960s and early 1970s was manifested in other ways. In 1970 concern over lax rules for controlling hazards in the work place led Congress to pass the Occupational Health and Safety Act. Under the provisions of this act the regulation of on-the-job RF hazards was split between two agencies: the Department of Labor's newly created Occupational Safety and Health Administration (OSHA) and a newly created agency in HEW, the National Institute for Occupational Safety and Health (NIOSH). NIOSH's main task in the RF area was to assemble information on bioeffects that could be used by OSHA for setting work place standards.

OSHA was empowered to set standards or, as an interim measure prior to the evolution of new standards, to adopt existing government or private standards.[15]

There was also concern over the degree to which the public as a whole was protected from general environmental hazards, such as those posed by the proliferation of communication equipment throughout the country. Every time a new microwave relay or radio broadcasting tower was erected, public exposure to RF radiation increased. On July 9, 1970, a reorganization plan signed by President Nixon centralized responsibility for pollution in the newly created Environmental Protection Agency (EPA).

EPA's responsibilities in the RF area filtered down through its complex bureaucratic structure and lodged in three offices. The Office of Radiation Programs (under the assistant administrator for air, noise, and radiation) took on the task of determining how much RF radiation is present in the environment. The Office of Research and Development, located at the Research Triangle in North Carolina, became the focus of EPA's basic research program. Its primary mission was to provide information on bioeffects that could be used by a third office, the Office of Planning and Evaluation, to draw up and have approved guidelines for regulating general population exposure to RF radiation. Such guidelines when formalized were to be forwarded to the president, who could use them to regulate government policy.

With the creation of EPA, the mechanisms were in place to deal with the RF bioeffects problem at its three most obvious points of public contact: the Bureau of Radiological Health (BRH) in HEW controlled RF radiation at its source (electronic products), OSHA (with the help of NIOSH) took care of the work place, and EPA oversaw the environment in general. If each of the designated branches of the executive fully exercised the authority given to it by Congress, there should have been very little chance of anyone, either at work or in the home, coming in contact with potentially harmful doses of RF radiation. But the solution to the RF bioeffects problem turned out to be more elusive than Congress foresaw. The legislative fiats demanding public protection in practice had very little impact on the actions government took through the executive to protect the public from the potential hazards of RF radiation.

Executive Actions

By the late 1960s one major piece of RF technology stood out above all others as a target for regulation, the microwave oven. Following the initial raising of consciousness about potential radiation problems, numerous checks were run on microwave ovens. The results suggested possible problems. A spokesman from Walter Reed told Senate committee members that twenty-six of thirty new microwave ovens purchased by the army leaked excessive amounts of radiation. One older oven was found to emit "200 mW/cm^2 out to a distance of 18 inches. . . . This is with the door supposedly closed."[16] On May 23, 1969, New York City's director of radiation control, Hanson Blatz, testified in Congress that seven of eleven ovens just purchased by the city did not meet the consensus standard generally followed at the time. In January 1970 random surveys conducted by HEW "had shown that one in three leaked enough radiation to present a health hazard to users."[17] Surveys begun in 1963 at the University of Washington at Seattle consistently found over the years that some ovens leaked radiation near or above the $10mW/cm^2$ cut-off point for presumed safety. Similar surveys in Pennsylvania reported that 16 percent of the ovens tested leaked more than $10mW/cm^2$.[18]

Although such leakage generally was thought not to present any significant hazards for occasional users, there were concerns. Microwave ovens were still a novelty in the late 1960s and therefore sometimes subjected to unconventional use. The University of Washington studies caught many people placing their faces directly against the screen of units in the hospital cafeteria while the unit was operating. One technician tried "to dry a vacuum tube in one of the ovens and in the process the tube erupted, spraying finely divided glass particles over the area."[19] The uncertainties brought about by such reports pointed to the need for executive action.

Hammering out a microwave oven performance standard took considerable negotiation. Industry was convinced that even if their ovens did on occasion emit higher than anticipated levels of radiation, the resulting exposures had not caused any RF-related health problems. Raytheon's spokesman, John Osepchuk, argued at the time and has continued to maintain to the present "that there never has been any verified case of injury to a human due to exposure from microwave leakage from microwave ovens.[20] If this were the case, then any call for stringent regulation was unnecessary. Government, on the other hand, knew that its surveys and scientific knowledge were not complete. True, there may not have been any verified reports of injuries, but there were no exhaustive surveys looking for injuries. Given the uncertainty that existed and the public concern, the agency that had been authorized to regulate electronic products, BRH, decided to act.

The microwave oven standard that BRH adopted in October 1970 reflected compromise figures advanced during the 1967–1968 congressional hearings and subsequently formalized by a fifteen-member committee. Industry representatives were comfortable with the military's and ANSI's $10mW/cm^2$ standard. They were willing to design ovens so that the radiation leaked fell within this limit, even at very small distances. (The limit for measuring closeness was eventually set at 5 centimeters since this is "as close as a human eyeball can come to the oven."[21] The eye was of crucial importance since cataracts were still thought to be a possible health problem.) BRH basically agreed with this reasoning but wanted additional conservatism, achieved by lowering the numbers slightly, allowing microwave ovens to leak $1mW/cm^2$ at 5 centimeters when new and $5mW/cm^2$ after being in operation. This performance standard became effective on October 6, 1971, one year after its publication.[22]

BRH's microwave oven performance standard provisionally set a very conservative trend for government regulation. The base figures used (1 and 5 mW/cm^2) fell below the commonly accepted $10mW/cm^2$ ANSI and military standards. More important was the fact that these lower limits were being measured very close to the oven itself. Since actual exposure levels decrease dramatically with distance (doubling distance decreases exposure four times, a fourfold increase in distance decreases exposure sixteen times, and so on), the BRH microwave oven performance standard gave the average member of the public much more protection then would be gained from allowing actual human exposure to approach $10mW/cm^2$. If other standards or regulations followed this trend, a comprehensive program for protecting the public from potential hazards could readily have fallen into place.

BRH's microwave oven performance standard did not provide a portent of things to come, however. Its conservative, interventionist approach to the RF bioeffects problem has not been duplicated by other executive actions. In fact, as of early 1984, it is impossible to cite any executive or congressional action, besides the microwave oven performance standard, that has established a firm federal policy for regulating exposure to RF radiation.

Following publication of the microwave oven standard, BRH continued to survey the performance of other devices that emit RF radiation with an eye toward regulation. In 1972 BRH carried out environmental monitoring in the Las Vegas and Washington, D.C., areas to find out how much RF radiation radio broadcasting stations and other sources emit. By 1975 reports on diathermy equipment began to appear, followed a year later by a preliminary report on marine radar systems. In 1979 CB radios were added to the list of possible sources of harmful doses of RF radiation, followed in 1980 by electronic security systems. The most recent additions to the list of equipment in possible need of regulation have been the RF sealer and video display terminals (VDTs), which comprised the subject of a House inquiry in May 1981.[23]

None of these initiatives has led to the adoption of additional emission standards. BRH's environmental monitoring program was taken over by EPA, which has yet to lead to constructive ends. The Federal Communications Commission has considered taking over responsibility for radio broadcasting stations if some way around the standards question can be found. RF sealers are generally thought to comprise a significant problem, but there are not enough new ones produced each year to warrant emission standards. Controlling old RF sealers falls within the domain of OSHA, which has been unable to come up with work place standards. VDTs remain controversial and unregulated, as do citizen band radios. Although BRH has tried on numerous occasions, it has not been able to mount a campaign to see any other emission standards through the regulatory process.

As weak as BRH's regulatory record has been, it has not been surpassed by the performance of other regulatory agencies in the federal government. OSHA began an aggressive campaign to regulate work place exposure shortly after it was created in 1970. Acting under the authority of the Occupational Safety and Health Act of 1970, it recognized ANSI's 1966 standard as a national consensus standard and adopted it as OSHA's own temporary radiation protection guide. The 1970 act gave

OSHA permission to adopt national consensus standards without prior notices, hearings, and reviews. The formal proceedings were to follow when OSHA abandoned the temporary consensus standard for its own work place standard, which would be developed through normal regulatory procedures.

OSHA's progress toward its own standard received its first setback in 1975. By this time OSHA had decided that the adopted ANSI standard was adequate for regulating work place exposure and therefore abandoned any attempt to develop its own standard.[24] The adopted ANSI standard proved to be unenforceable, however. OSHA discovered this fact when it cited the Swimline Corporation for violating the standard.

A check of Swimline's plant in Commack, New York, on February 27, 1975, revealed that an employee working at an RF sealer was being exposed to more than 10mW/cm^2 of RF radiation. A citation was issued on March 10. Swimline contested the citation, arguing that the standard it was reportedly violating was advisory, not mandatory, and therefore Swimline could not be held accountable. An administrative law judge agreed with Swimline and essentially threw the 10mW/cm^2 standard out: "The standard set forth . . . is not an occupational safety and health standard as defined by . . . the [Occupational Safety and Health] Act."[25]

The judge's decision made no pronouncement about 10mW/cm^2 as a presumed safe level for exposure to RF radiation. It simply noted that because the phrase "should not be exceeded," rather than "may not" or "will not," was used in framing ANSI C95.1, OSHA had no right to adopt this standard as mandatory, and employers could not be cited for violating it. In other words, as of December 31, 1975, when the Swimline Corporation decision was issued, OSHA had no rule for protecting a worker at an RF sealer from potentially hazardous doses of RF radiation. Workers could be, and continue to be, exposed to levels of RF radiation that exceed the ANSI standard, that fall into the range of known thermal effects, and that have not been extensively studied for possible long-term athermal effects.[26]

OSHA initially seemed bent on ignoring the Swimline decision. In testimony submitted to a Senate committee in 1977, the director of OSHA's Health Standards Program, Grover Wrenn, contended that " 'should' standards are mandatory," the judges' ruling notwithstanding. The justification for this position was the contention that employers violating the 10mW/

cm^2 standard had to show that "careful consideration" had been used in making the decision to exceed recognized safe levels. Had Swimline "failed to follow the radiation protection guide without giving 'careful consideration,' a citation for violating [the guide] would have been proper." Whether any administrative law judge would have agreed with this position is doubtful. It is even questionable whether OSHA took it seriously. OSHA submitted no written briefs in the 1975 Swimline case, did not contest the ruling in the case when it was reaffirmed in 1977, and eventually reluctantly pursued the only other course of action open, setting its own RF standard.[27]

The normal course of events that OSHA follows in setting standards is to turn first to NIOSH for technical advice. NIOSH responds to requests for advice by instituting reviews that result in technical surveys known as criteria documents. The criteria documents, which can contain recommendations for standards, are forwarded to OSHA for action. After reviewing a criteria document, OSHA can accept NIOSH's recommendation, modify it, or reject it, depending on the understanding of its effect on work place conditions and the economy.

OSHA-NIOSH deliberations on the RF problem began on a shaky footing in late 1977. By this time NIOSH had been looking into possible RF problems in the work place and had already issued one provocative report on RF sealers, which suggested that as many as three in every four RF sealer operators were being exposed to excessive (above $10mW/cm^2$) amounts of RF radiation.[28] When asked about this report in 1977, OSHA responded that it had "received no information from . . . NIOSH to indicate that there are adverse health effects associated with the use of radio frequency radiation equipment at the levels of exposure presently encountered." By this time OSHA's own interim standard had been rejected twice in the courts, and RF bioeffects were beginning to become a topic of nationwide interest. Thus by June 1977 OSHA had finally "requested that NIOSH provide us with recommendations, perhaps in the form of a criteria document."[29]

NIOSH responded by formally undertaking the development of an RF criteria document in May 1978. The initial target date set for the final document was early 1980. Most of the work would eventually be done by the Special Hazards Review Program and its director, Zory Glaser. Glaser had been involved with RF bioeffects in one way or another since his navy days in the late 1960s and had become well known in the field for his annual comprehensive reports on RF bioeffects literature. Under his supervision the projected document grew into a massive two-volume survey that attempted to base recommendations for standards on the latest and most complete scientific information.[30]

By the time the criteria document started making rounds among external reviewers in May 1979, it had become apparent that basing recommendations for a work place standard on objective interpretations of the scientific literature was going to be difficult, if not impossible. Supporters of the commonly accepted 10mW/cm^2 standard urged NIOSH to exclude all controversial and unsubstantiated scientific deliberations from the criteria document. Critics of the old 10mW/cm^2 standard strongly urged that all such literature, however controversial, needed to be considered. In a long, bitter letter to NIOSH director Anthony Robbins, Louis Slessin of the Natural Resources Defense Council outlined the shortcomings of the criteria document that were apparent from the public's point of view. He also noted how difficult it had been to get a copy of the draft document even though he had been promised a copy as soon as it became available for external review. The ultimate objective of Slessin's letter was clear; it was intended to urge NIOSH "to withdraw the draft Criteria Document on Radiofrequency and Microwave Radiation, and to commission a new occupational health study."[31]

In response to such pressure, NIOSH officials followed half of Slessin's advice; they withdrew the draft document from consideration. It has not yet been replaced. After 1980 NIOSH's RF bioeffects initiative floundered. Discouraged by his dealings with an uncooperative bureaucracy in NIOSH, Glaser quit his job and moved to BRH. His replacement, Paul Strudler, fought to keep the RF criteria document alive but without much success. The change in attitude toward standards that came in with the Reagan administration further eroded interest in having such a criteria document prepared. The result of the failure of NIOSH to produce the long-promised document has been that OSHA has not received the advice it sought and has not imposed an RF standard for the work place. The congressional mandate to provide for the safety and health of U.S. workers set forth in the 1970 Occupational Safety and

Health Act has yet to bear any lasting fruit in the RF bioeffects area. The same conclusion follows for environmental exposure.

EPA devoted most of its concern over radiation problems in the early 1970s to the ionizing portion of the spectrum. No formal programs in the nonionizing field had been initiated by EPA when officials were called to testify before the 1973 Senate Commerce Committee hearings on PL 90-602. Four years later, when the Senate Commerce Committee again convened hearings on PL 90-602, this situation had changed dramatically. Between 1974 and 1977 EPA instituted two major research efforts designed to assemble enough information to establish a general environmental RF standard if one were needed.

One program, for environmental monitoring, was designed to determine how much RF radiation was present at select locations throughout the country. Beginning in October 1975 a specially outfitted van toured twelve major urban areas, from Boston to Portland, measuring ambient levels and pinpointing sources. Potentially high-exposure areas (relative to the population as a whole) were targeted for special measurement programs. Equipment was hauled to the top of the World Trade Center in New York, the Sears Tower in Chicago, and the Home Tower in San Diego to measure how much radiation local radio and television transmitters were inflicting on upper-story office workers. When this information was assembled, it provided EPA officials with a clear picture of the actual exposure levels that needed to be reviewed for potentially hazardous effects. These surveys showed that less than 1 percent of the U.S. population routinely was exposed to more than 1 $\mu W/cm^2$ of RF radiation. Persons in high-exposure areas—the tops of buildings with clusters of RF transmitters—could be exposed to levels that ran as high as 100 to 200 $\mu W/cm^2$.[32]

EPA's second research program, developed under the supervision of Dan Cahill at the Research Triangle facility in North Carolina, was intended to provide scientific data on the levels at which biological effects might be expected. In reviewing the bioeffects question, EPA cast its nets broadly, even duplicating some Soviet experiments, to determine whether it was possible to get the same results. (In some cases it was.) Long-term plans then called for combining the bioeffects data with the survey data to produce recommendations for a general population exposure standard.

In 1977 EPA projected that key decisions on an RF standard would be made in 1978. EPA officials testified in June 1977 that "sometime next year, hopefully by March of 1978, we expect to have a fix on what the health effects are and a fix on what the levels of the environment are. Based on this assessment of the scope of the problem, and then using our Federal guidance authority, we hope to promulgate guidance in the area for a general population standard and perhaps even to go into other areas."[33] By July 1979 the target dates had been moved back: "ORP plans to prepare an Initial Draft of the Proposed Rulemaking by April 1980 and a final Draft by March 1981."[34] As these dates came and passed, EPA along with the other regulatory agencies entered into a period of contracting budgets and uncertain commitments. The research budgets for the Division of Experimental Biology were cut and the monitoring program reduced. In April 1981 the target date for the first draft of the notice of intended rule making was moved to December 1981, with the final draft pushed off to December 1982.[35] When the December 1981 date rolled around, the decision on whether to proceed was further delayed; it is still being delayed as of early 1984.

With EPA's failure to regulate, the final piece in the intended three-part protection plan was lost. With the exception of microwave ovens, the public is not protected by the federal government from potentially harmful emissions from electronic products. There are no RF work place standards or general population exposure standards.

Question of Need

Government officials have asked for advice on the underlying question of need on many occasions. They have not automatically assumed that the authority to regulate implies a mandate to regulate. In part this advice has been sought because it is required by law. The first step in rule making is to publish a notice of intended rule making, which alerts anyone who may be affected to the fact that government action is being contemplated. When such notices have been published or when Congress has held hearings to determine the need for enabling legislation, many opinions have been expressed.

Until fairly recently industrial spokespersons tended to urge government to go slowly in developing standards. When the possibility of regulation emerged in the late 1960s, industry worried about possible economic hardships that could result if standards were tied to product recalls, plant inspections, testing and certification, and seizure.[36] Similar sentiments emerged from industry a decade later when the FCC sought advice through a Notice of Inquiry on its role in regulating RF exposure. When questions about possible FCC involvement in monitoring standards arose, reservations were expressed. Most felt that industry, not the FCC, was better qualified to monitor exposure.[37]

In contrast, many members of the public have urged government to take an active role in regulating RF exposure and have been frustrated by the inability of government to do so. A long analysis of the RF bioeffects problem published in 1979 by Karen Massey of the Natural Resources Defense Council (NRDC) listed many areas in which government could take a more active role in regulating RF exposure.[38] NRDC's speaker before the House Committee on Science and Technology hearings in 1979, Louis Slessin, expressed the same opinion. While not supporting a proposal being discussed at the time that would have seen Congress step in and set an RF standard, Slessin did recommend that "Congress . . . tell the agencies to get cracking and . . . set a standard."[39]

This interventionist point of view, which until the late 1970s was held primarily by persons claiming to represent the public, has recently been picked up and echoed by some persons in industry. Industry frequently uses federal rules to prove to the public that its activities do not pose hazards. If the government has no rules, as is the case with RF radiation, industry can be left in an awkward position. This is why at the same time NRDC's Slessin advised Congress to urge the agencies to "get cracking," a spokesman for RCA, Howard Johnson, also pushed for action, asking Congress to adopt the ANSI standard as "an interim, mandatory, Federal standard. Once this stake is driven," Johnson continued, "we will have a rallying point from which research and regulation can progress in a more orderly and less-pressured fashion without unreasonable risk to the public."[40]

The change in industry's attitude toward federal regulation suggests that the need for standards is relative. Obviously their primary purpose is to protect. In this capacity they serve the need of the public; they protect the public from potentially hazardous environments or activities. But standards also supposedly establish an objective, consensus position on how adequate protection can be achieved. In this capacity they serve the needs of industry and law; they establish guidelines that can be used for judging the safety of proposed technological projects and for dealing with questions of liability in damage suits. Establishing standards also provides evidence for active intervention in problem areas, evidence that government needs when it is asked what it has done. Any one of these lines of reasoning could be advanced to argue that there has been and continues to be a need for RF standards. Congress's call for action over a decade ago seems not to have to have been misguided. But if the call for action was not misguided, why has so little been accomplished?

Shortcomings of the Regulatory Process

Over the past thirty years, two patterns of decision making have been used to tackle the RF bioeffects problem: centralized and decentralized. During the Tri-Service era the military relied on a centralized committee system to answer questions about bioeffects and set standards. Centralization and limited participation also characterizes the approach ANSI has used to handle the bioeffects issue.

In contrast the federal government's approach to the RF bioeffects problem can best be characterized as decentralized. Government decision making begins with elected representatives who identify critical issues, draw up legislation, and then hand problems over to the executive for administrative action. In the executive, environmental problems are fragmented into many subproblems, which slowly filter down through bureaucratic channels until they come to rest on the desk of individual administrators, directors, and their staffs. The administrators or directors then proceed to resolve their assigned subproblems. Only after all of the individual subproblems have been solved can the entire package be reassembled and a solution to the complete problem achieved.

The major advantage of centralized decision making is that it gets things done. The military and ANSI succeeded where government failed. They followed the deliberation process through to final decisions and policy pronouncements. The major disadvantage of the centralized approach is that it is not sensitive to contrasting points of view.

The major strength of decentralized problem solving is that

it ensures maximum input into decision making. Different agencies in the federal government have different constituencies to satisfy. OSHA, in the Department of Labor, approaches problems from an industry-labor point of view, leaning more toward one or the other in accordance with the political philosophy of the administration in office. EPA brings in and, again depending on administration leanings, is more or less responsive to public and environmental concerns. NIOSH has other interests to serve, most of which tie closely to medical and scientific activities because of its research focus. The other federal agencies that have dealt with the RF bioeffects problem bring in other points of view, and there is the assumed direct link with the public through Congress.

The primary disadvantage of decentralized problem solving is more than evident in the failure of government to make major policy decisions about RF bioeffects. Spreading authority weakens mandates and clouds responsibilities. The consequence of no one being fully responsible has been that there is always someone else to turn to for answers or for assigning blame. As a consequence decentralized authority greatly facilitates buck passing and foot dragging.

Foot dragging is most problematic in the executive regulatory agencies. The real problem in this case is not so much foot dragging as foot weariness. The steps required to go from a perceived need to a standard are sufficient to exhaust even the sturdiest bureaucrat. EPA's early estimates for successful rule making in the RF bioeffects area ranged from three to five years, if everything went according to plan. Even the simplest decisions along the way can result in long delays. EPA took more than five years to decide whether the agency ought to announce that it was thinking of trying to generate an RF standard (publish a notice of intended rule making). Before this decision was made, agency administrators all the way up to EPA's director had to be convinced that there was a need, that the need outweighed potential costs in terms of actual dollars and political considerations, and that there was sufficient urgency to warrant allocating scarce agency funds to the pending project.[41]

The debate that goes on within the agency at this early point in rule making is between the RF bioeffects experts who in some cases have been working on this problem for over a decade and their administrative superiors. The on-line experts in

EPA in the early 1980s, such as the director of the research program, Dan Cahill, the head of measurements, Rick Tell, or the manager of the standard effort, Dave Janes, draw the evidence together and make their case to their administrative superiors. The upper-level bureaucrats look at the evidence, weigh it against other pressing needs, test the political and philosophical waters, and make the decision or simply let the material sit on their desks. During this process even factual reports can meet with delay. Cahill spent hours pouring through the bioeffects literature and drawing together a comprehensive survey that could be used for decision making. His literature survey became bogged down in the in-house review process. Cahill is now working for private industry.[42]

The list of such stories could be multiplied. When NIOSH's new director decided in 1981–1982 to disband the Washington facility and consolidate operations in Cincinnati and his home state of Georgia, NIOSH officials working on the criteria document spent time putting their houses up for sale and playing computer games with the Department's unused search equipment. BRH has watched hundreds of hours of work on diathermy regulations scrapped because of uncertain priorities. Simply getting approval to publish a paper or give a talk, which will be prefaced by the usual disclaimer, "personal comments and not agency policy," can take months.

The decentralized decision-making process used to protect the public and set standards is an obstacle to effective and efficient regulation. This fact has caused more than one observer to despair, as did Congressman Jerome Ambro in 1979 when he tried to tackle the microwave debate: "There is no question that in looking at some of the recent foot dragging by the FDA that one is less than sanguine about your ability to act in less than 5 to 8 years on anything, from a determination of whether something is carcinogenic to how much salt should be allowed in children's food." This was the opinion of someone who believed that "without independent regulatory agencies the United States would have been raped a long time ago by those groups that the agencies were set up to control."[43]

Many steps have been taken to get past the obstacles presented by the regulatory bureaucracy. In the late 1960s a group of concerned administrators and scientists formed an advisory committee to coordinate RF bioeffects work. The committee, known as the Electromagnetic Radiation Management Advisory Committee, initially had the prestige of a White House office behind it, the Office of Telecommunication Management. Under President Carter the committee was moved deeper into the bureaucracy of the executive when it was reorganized out of the White House and into the National Telecommunications and Information Administration (NTIA) in the Department of Commerce.[44] Under President Reagan it was phased out as part of a program to reduce federal spending and limit regulatory activities.[45]

Another effort to overcome bureaucratic obstacles began in 1978 when an interagency committee on regulation, the Interagency Regulatory Liaison Group, voted to establish the Radiofrequency and Microwave Committee. The objective of the new committee was to ensure that its members—the Consumer Product Safety Commission, the Department of Agriculture, EPA, FDA, OSHA, and NIOSH—pursued a unified approach to regulation. A year later NTIA responded to a request by Frank Press, President Carter's science adviser, for "a detailed plan for a federal program on understanding the biological effects of nonionizing electromagnetic radiation," by establishing the "biological effects of nonionizing electromagnetic radiation" or "BENER" task force.[46] The BENER task force issued one report, which has gone the way of all other reports in this area and been ignored; the radiation committee has been dissolved.

Time and again the members of the agencies who deal directly with the RF bioeffects problem have recognized that cooperation is the key to success: "Let us have NTIA provide a home and some small staff support to the interagency coordinated program for the health effects, and have as the main driver, the group that is responsible for making this Federal program work; this coordination council [should be] made up of the line managers from each of the agencies."[47] The managers know what has to be done, but they have never been allowed to work together for very long. There seems to be no way to overcome the negative forces of decentralization and to tap the rich reservoir of information and expertise that the branches of the executive have developed since the late 1960s.

The problem of decentralization could be solved by Congress, which could mandate that specific objectives be met within designated time frames, but this is where the other negative attribute of the decentralized system comes into play: buck passing. Congress may have intended to find long-term, far-reaching solutions to the RF bioeffects problem when it became public in 1967, but it has never gotten beyond crisis-to-crisis stopgap measures. GE's misdesigned television sets first brought the RF problem to the attention of Congress. By 1973 the focus shifted to microwave ovens. The day before Senator John Tunney of California opened a second round of Commerce Committee hearings on the radiation control act, *Consumer Reports* stamped "not recommended" on microwave ovens. Tunney and his colleagues wanted to know why, given the fact that BRH had supposedly solved this problem. Four years later the Commerce Committee was back in action, this time under the guidance of Senators Wendell Ford and Warren Magnuson, asking the same questions about RF radiation, now in response to the Moscow embassy crisis and a not-inconsequential related development, the appearance of Paul Brodeur's *New Yorker* articles in December 1976. House hearings on the solar power satellite project in 1979 probed into RF bioeffects because of the potential obstacle they presented to using microwaves to beam solar energy to earth-based power stations. Three months later a second set of House hearings on nonionizing radiation was heavily influenced by the interests of Congresswoman Elizabeth Holtzman, whose constituents in New York City were in the middle of a bitter debate aimed at setting a fairly stringent RF standard. More recently another set of House hearings, chaired by Albert Gore of Tennessee in May 1981, responded to the problem of VDTs and what one representative termed the "frankly shocking" issue of RF sealers.[48]

Although Congress has returned again and again to the RF bioeffects problem, it has never done so for fundamental reasons; more immediate problems have been on the collective congressional mind. And once the immediate problems have been addressed in some way the need to probe more deeply and broach genuine long-term solutions disappears. In 1968 passing the RF problem along to the secretary of HEW got rid of it. Consciousness raising through the hearing process seemed sufficient to alleviate the concerns that arose in Congress in 1973 and will probably be the end of the 1981 congressional worries over RF sealers. The 1977 concerns over the Moscow embassy affair and Brodeur's articles were satisfied by issuing two reports: one reiterating the need for further research, the other mildly critical of the handling of the Moscow embassy affair.

The here-and-now concerns of Congress are particularly apparent in one sustained effort it undertook to solve the RF bioeffects problem. In casting about for solutions for the energy crisis in the 1970s, one potentially fruitful scheme that emerged was the idea of collecting free, solar energy in space, concentrating it, and beaming it back to earth with microwaves, which could be converted to electricity. The solar power satellite project, as it came to be known, raised numerous environmental concerns. Getting enough equipment into space to collect the amounts of energy needed to reduce U.S. dependence on foreign oil required hundreds of trips into space, which in turn would cause atmospheric pollution. Once the equipment was in space and operational, the question of whether the microwave beams being transmitted back to earth would have biological effects became important. When Congress asked the experts whether the bioeffects problem was potentially serious, they were told that no one could provide a simple answer to this question.

Uncertainty in this case frustrated congressional members. They wanted to proceed with the satellite project and needed answers immediately. Thus they introduced legislation in the Ninety-sixth Congress that proposed to spend $25 million getting answers.[49] In March 1979 the head of NASA, William Frosch, told Congress that more funds would not and could not get faster results. Frosch strongly opposed pursuing a "shotgun approach of just tossing tens of millions of dollars so everybody can go out and research his favorite subject." Eight million dollars was at the time sufficient for a careful, phased preliminary assessment of the satellite project; the $25 million Congress suggested was not needed. This was not the answer Congress wanted. Frosch was proposing "to proceed much slower than this committee would like for you to proceed."[50] Accordingly the congressional prodding continued, despite the advice it received to the contrary. On January 5, 1981, the Solar Power Satellite Research, Development and Evaluation Program Act was reintroduced into the Ninety-seventh Congress, having failed to make it through previously.[51]

The immediate needs precipitated by the satellite project came to an end when subsequent review showed that it embodied more problems than initially anticipated. As a result the idea of beaming energy back from space has been abandoned for the present, which means that Congress no longer seems to need an immediate, long-term solution to the RF bioeffects problem. When uncertainty stood in the way of technological development, there were demands for action. Health concerns seem not to provoke the same urgency as possible delays in technological expansion.

7

Science, Scientists, and Science Policy

The lesson is that we must evolve methods of managing and coordinating research without inhibiting the investigator's freedom. . . . [This] is of particular importance when basic research pertains to a matter of public interest. You must have independent research if you expect to get results that can be believed. At the same time you have to be able to channel the efforts in a direction most likely to produce the necessary results.—Charles Süsskind, Richmond Symposium, 1969

Shortly after it was delegated regulatory responsibilities in 1968, BRH asked a researcher at nearby Virginia Commonwealth University, Stephen Cleary, to help assemble a group of scientists who could assess the present state of knowledge on RF bioeffects. BRH's expectations at the time were clear: "We are . . . very much interested in the total contribution which this Symposium will make to our present knowledge and the guidance you will offer us for future efforts in developing required performance standards for devices that emit unneeded microwave radiation."[1] BRH officials felt that the scientific community could contribute "important value judgments which will make a positive impact on the activities" of government in the "day-to-day administration of . . . Public Law 90-602."[2]

Scientists traditionally have responded eagerly to government calls for assistance. Since providing advice and receiving support customarily go hand in hand, scientists have been more than willing to help government solve its problems. This was particularly true in the late 1960s. Even before PL 90-602 was passed, former Tri-Service researchers used the occasion of the completion of the last Tri-Service project, awarded to Michaelson's group at Rochester University, to convene a panel discussion on the "Biological Effects of Microwaves: Future Research Directions" at the 1968 International Microwave Power Institute annual meetings.[3] With two consenting partners, it should come as no surprise that RF bioeffects policy of the last decade and a half has been primarily the product of a marriage between scientific research and government (including the military) support.

Objectives versus Priorities

The initial response of the technical experts to the perceived crisis of the late 1960s was mixed. The relaxed attitude of industry was based heavily on years of experience with RF technology, coupled with a considerable dose of skepticism about experiments that pointed to possible low-level effects. This position was succinctly argued in 1968 by an engineer from the Chemetron Corporation: "I started in with a 50-kW transmitter at 19, unshielded and not very far from the antenna. Apart from losing my hair I feel pretty good. I wonder how or what you have been injecting into the organisms so you get these effects?"[4] Many experts in the late 1960s were comfortable with the overall conclusions drawn from the Tri-Service program and were willing to trust to industry and to the military, the two primary users of RF technology, the task of monitoring bioeffects.

This view, which was principally maintained by persons (not all) in industry and the military, was vigorously contested by most university and government scientists. From a purely scientific standpoint, the research of the 1950s had barely scratched the surface. Researchers had not succeeded in standardizing experimental techniques when the Tri-Service funding ended. They had not looked in depth at special exposure situations, such as pulsed power. This left researchers such as Tufts University eye specialist Russell Carpenter skeptical about past policy decisions: "I am not ready to be comfortable in a microwave field when I am told that the average power is safely below $10 mW/cm^2$ but where I am being subjected to peak powers many times that figure. I would prefer not to be there."[5] The obvious remedy for this situation was more scientific research.

Most scientists who favored more research were willing to

cast their nets broadly. John Howland, who worked with Sol Michaelson, suggested that serious attention be given to "numerous curiosities" such as "the howling of dogs near transmitters, peculiar actions of birds, fatigue and headaches in workers, and other psychosomatic complaints."[6] When Howland made this recommendation to colleagues in 1968, the chances of pursuing such research looked brighter than they had a year or two before. The pending radiation control legislation suggested that "we are going to see some interest on the part of the federal government in electromagnetic radiation." Interest was an obvious step toward renewed support. Former Tri-Service researcher Charles Süsskind from the University of California, Berkeley, looked for this renewed support "in the very near future."[7] His prediction proved correct. Shortly after PL 90-602 was passed, funding for RF bioeffects research became available again.

In its rush to get legislation on the books to deal with the perceived environmental crisis of the late 1960s, Congress by and large ignored organizational problems. Many of the scientists who stood to benefit from the research funds that were bound to follow understood that this situation was not satisfactory. Accordingly when the Office of Telecommunications Policy let it be known that it might be interested in helping to coordinate RF bioeffects research, the scientific community responded quickly. By late 1968 Pandora adviser Sam Koslov had gathered together a few colleagues and drawn up plans for an Electromagnetic Radiation Management Advisory Council (ERMAC). Within a few months his plans had been accepted, and ERMAC's first official meeting was convened in March 1969. The principal agenda item at this and most subsequent meetings was the formulation of recommendations for a coordinated RF bioeffects research program.

ERMAC's members wasted little time getting to the heart of the RF bioeffects problem: "There appeared to be general agreement that the first order of business should be the study of the $10 mW/cm^2$ radiation limit established within the U.S. as compared with the $10 \mu W/cm^2$ limitation established by the Soviet Union." By the third meeting (June 5, 1969), Koslov was assigned the task of preparing a draft of a document describing ERMAC's aims and objectives. His first draft "generated considerable discussion with resulting constructive comments." A second draft presented to the fifth ERMAC meeting in September 1969 also provoked discussion, and it was completely revised again. Koslov's description of ERMAC's aims and objectives by this time had become a proposal for a coordinated research program. After additional meetings and many exchanges by letter, ERMAC's aims and objectives were finally printed in December 1971 as a *Program for Control of Electromagnetic Pollution of the Environment: The Assessment of Biological Hazards of Nonionizing Electromagnetic Radiation*.[8]

The problem-solving framework ERMAC's program established was unquestionably scientific. Its opening paragraphs juxtaposed a compelling discussion of the problem against "the solution—a research program." The need for more scientific research was sketched out in unmincing terms: "Unless adequate monitoring and control based on a fundamental understanding of biological effects are instituted in the near future, in the decades ahead, man may enter an era of energy pollution of the environment comparable to the chemical pollution of today." Critical areas were singled out for special attention. It was argued that "the consequences of undervaluing or misjudging the biological effects of long-term, low-level exposure could become a critical problem for the public health, especially if genetic, clinical, physiological, and behavioral effects of electromagnetic radiation at power densities below" $10 mW/cm^2$. The only way out of the dilemma posed by this situation was more research.[9]

From the need for more research, ERMAC's members turned to specific objectives, identifying four areas as demanding the most attention: long-term, low-level studies; epidemiology; research on basic mechanisms; and coordination activities. In combination these four areas covered the main elements of RF bioeffects research, thereby ensuring that a wide variety of projects would be covered. The program was projected to cost $63,435,000 for the first five years.

Similar broad recommendations emerged from other quarters in the late 1960s and early 1970s. One industrial representative who favored more research, Bell Labs's George Wilkening, summarized the feelings of his colleagues at the 1969 Richmond symposium: "there seems to be unanimous or almost unanimous agreement that one of the things that should be done is to perform repeat insult type experiments at low levels."[10] A simultaneous plea for more epidemiological data was made by Norman Telles of BRH, who estimated that U.S.

researchers had conducted only one extensive eye study and three lesser general surveys that covered fewer than 400 individuals who had had on an average less than three years of exposure.[11] Coordination and information on basic mechanisms were universally agreed to be essential. In sum ERMAC's program succeeded in capturing the feelings of the scientific community and in translating them into a document that could be used to establish a broadly based scientific research program on RF bioeffects.

The broad, ideal objectives established in ERMAC's program did not fare well when it came to establishing priorities. Although most researchers agreed in principle that more attention had to be paid to epidemiology, long-term low-dose studies, and the like, they seldom were willing to assign high enough priorities to such studies to get them funded. When push came to shove, the scientists who advised on RF bioeffects research retreated to the attitudes that had led to the dominance of thermal thinking in previous decades. These scientists were at home with precise, controlled experiments; they were not comfortable with effects that lacked causal explanations. This inevitably narrowed the focus of the RF bioeffects research to controlled animal experiments, theoretical modeling, and a concentration on thermal effects. The old way of thinking died hard.

A researcher who typifies this mode of thinking is Sol Michaelson, who has been a constant participant in the advisory process. Like the rest of his colleagues, he was willing to sketch out broad research agendas in the late 1960s to get research funds flowing again. But his underlying formula for judging the best research remained selective: "The biological indicators of microwave response should be easily replicable and the technique of measuring should not require elaborate equipment. Microwave induced biological changes should have a high probability of occurring. The range of experimental values for the parameter selected should be well defined and should have a very poor range of variability. Ideally the range of normal values . . . should be the reference value when evaluating the pool exposure data."[12] Michaelson's thermal experiments provided the ideal model for this type of experiment. The variable—temperature change—was decisive, easily measurable, and relatively quick in occurring. Normal temperatures could be used for establishing a background against which change could be measured. Michaelson looked for similar crisp rigor in other RF bioeffects research. The degree to which experiments lived up to his expectations determined the extent to which he recommended their being supported.

The manner in which these attitudes came to dominate RF bioeffects research and policy is apparent in the activities of ANSI C95 after the adoption of C95.1 in late 1966. The framer of this first standard, Herman Schwan, began to relinquish control of ANSI activities in November 1965 when he resigned the chairmanship of C95. His place was temporarily filled by the C95 secretary, Glenn Heimer, until a new chairman, Saul Rosenthal of the Polytechnic Institute of Brooklyn, was appointed in June 1968.[13]

Rosenthal took over the chairmanship under difficult circumstances. At the time Schwan was on sabbatical in Germany and had left little information behind on how C95.IV had functioned. The records Rosenthal had were "not complete since there is a jump between 1961 and 1963 from a committee of 14 members headed by Colonel Knauf to a committee of 5 members headed by Dr. Schwan. In addition there seems to have been an involvement with the University of Miami and that seems to have disappeared also. In 1963 there appears some information on the proposed standard; however, there was no background information on how it was arrived at."[14] This situation was not rectified until the summer of 1970, when Schwan began attending C95 meetings again and set out for colleagues his account of the standard-setting process.[15] In the meantime Rosenthal had to begin reassessing the RF bioeffects problem, working under the deadline of ANSI's requirement that its standards be reevaluated every five years.

The initial reassessments undertaken during the first few years of Rosenthal's chairmanship singled out two key questions that needed to be addressed. First, the sketchy information available to C95 members left little doubt that C95.1 was based heavily, if not exclusively, on data collected prior to and during the Tri-Service era. One question that had to be answered was whether there was any new information that would require changing the standard. Second, even if no new data had surfaced, ANSI members still had to ask whether the data used to set C95.1 were adequate. If they were not, provisions would have to be made for undertaking additional research.[16]

ANSI thinking on both questions was sketched out briefly by

Rosenthal in a letter to BRH director John Villforth, commenting on BRH's proposed microwave oven standard. Villforth obviously needed to keep abreast of ANSI deliberations since conflicts between BRH and ANSI standards could cause problems. In reassuring Villforth that ANSI had not changed its mind and that the proposed oven standard seemed reasonable, Rosenthal predicted that "C95 will find that no additional information has become available since the present standard was adopted that would indicate a change is warranted." His answer to the first question was that scientific research had not discovered anything that invalidated the $10mW/cm^2$ standard. But that did not mean that the data base used in setting C95.1-1966 was adequate. Rosenthal characterized this standard, somewhat paradoxically, as "an excellent one [that] still leaves much to be desired" because its data base was "deplorable." The obvious conclusion that followed was that "unless there is a vigorous and active program of research directed toward obtaining the pertinent information," one could not be sure of the validity of the standard.[17]

To rectify this situation, ANSI, like ERMAC, began planning for future research from a broad-based perspective. Shortly after Arthur W. (Bill) Guy took over the chairmanship of C95.IV in July 1970, five study groups were set up "to identify and document the requirements for additional information needed to modify or improve present standards": Near-Zone Field Effects, chaired by John Osepchuk (Raytheon); Frequency Effects, chaired by Albert Kall (Ark Electronics) and Sidney Kessler (U.S. Information Agency); Low-level (Athermal) and Modulation Effects, chaired by Allan Frey (Randomline); Environment, chaired by Bill Mumford (Bell Telephone); and Population Groupings, chaired by William Mills (BRH).[18] The coverage being recommended was comprehensive, but the majority of ANSI members were not willing to take steps that would have provided incentives to expand the base of RF bioeffects research. Instead they retreated to the safe world of controlled experiments and thermal thinking, ignoring the consequences this had on their assumed philosophy of standard setting.

One person tried to change this situation, a figure who was himself becoming a source of controversy—Milton Zaret. In a brief but provocative open letter to ANSI members written in April 1970, Zaret recast the language and philosophy of C95.1

in an effort to place a greater burden of proof on those who believed it was adequate. Since general population studies had not been conducted, Zaret proposed that the standard state that "the recommendations are not intended to apply to the general public." Similarly he recommended that pulsed radiation with peak powers more than one hundred times their average and nonuniform fields be excluded, again reflecting the lack of data available. To ensure that other potential problems were not missed, Zaret suggested requiring that "when a radiation generating system either is capable of exceeding the recommendations or is not adequately defined by this guide, then . . . the user should ensure its safety by performing appropriate biological assay experiments." And to avoid providing an image of certainty where none existed, Zaret's final recommendation would have inserted probability into the entire standard by changing the phrase explaining the safety of below threshold exposures from "will not" to "is believed not to result in any noticeable effect to mankind."[19]

Had Zaret's proposed reworking of C95.1 been accepted, it would have changed ANSI's philosophy of standard setting and thereby the accepted protocol for RF bioeffects research. The military and other users would have been compelled to pursue general population studies or tell exposed populations that they did not have the evidence to guarantee safety. Industry would have had to run biological assay tests on RF equipment before subjecting workers to it, not after. ANSI would have had to be sure of its scientific information before issuing a firm standard. By shifting the burden of proof, Zaret's proposal would have made long-term, low-dose, and epidemiological studies a necessity.

Zaret fully understood the consequences of his suggestions: "The effect of the peak power reduction to a level justifiable by data would be simply to shift the responsibility for personnel safety to the user, that is, to the military; to *remove* the sanction the document now offers to the use of high pulsed power radars, and thus force the user to justify his safety codes of practice by supporting or engaging in appropriate research."[20] Industrial representatives, such as John Osepchuk and Paul Crapuchetts (Litton Industries) also understood the consequences of the proposed changes and argued vigorously against them: "Most military radars have duty cycles near .001, or peak powers near 1000 times average, so that the revised document would no longer apply. Also, if consumer products utilizing microwave power became widely used, the document would not apply there either. So where *would* it apply, and wouldn't this restrict the document into uselessness?"[21] Zaret was unmoved by such seemingly pragmatic arguments. If the military had to change its defense plans as a result, so be it; but necessity, at least not Zaret's version of necessity, did not rule the day.

Zaret's proposal for revising C95.1 was circulated, criticized, and rejected. In rejecting Zaret's suggestions, ANSI members permitted their recommended research program to remain squarely in the tradition set in the 1930s and followed after World War II. The final report of ANSI's study groups, "Research Needed for Setting of Realistic Safety Standards" (August 1, 1972), mentioned, but suggested no practical means for resolving, the old problem of occupational versus general population exposure.[22] The least developed sections of the report were those that dealt with population groupings and the environment. This all but assured that in the years to come, little serious attention would be given to epidemiological studies or long-term, low-level experiments. Instead the dominant notes struck in the study group report were measure, calculate, and replicate, mostly within the context of animal experiments and for the purpose of learning more about basic mechanisms. This is the course most RF bioeffects research would follow throughout the remainder of the 1970s and into the 1980s.

Inconsistencies and Biases

The emphasis placed on controlled animal experiment was not without justification. Knowledge of the precise mechanism for RF-tissue interaction could be useful when making decisions on exposure standards. This is precisely the sort of information BRH and others were seeking in the late 1960s when they turned to the scientific community for help.

The research plan adopted in the 1970s, however, had internal problems. To begin with, few persons asked whether the research goals that had been established were achievable. Time is needed to solve complex scientific problems. In this particular case the possibility existed that RF bioeffects might be dependent on frequency and on tissue type, thus opening the way for thousands of different interactions. In addition RF bioeffects researchers were studying operations within the body that themselves were not fully understood. Given these complications, there was no guarantee of success. The $63 million being asked of government could have been a down payment on a product that could not be delivered. Recognition of this possibility in the early 1970s might have prevented some of the apologizing that had to be done in the late 1970s when Congress began to wonder why the problem had not been solved.

More important, the product that the scientific community has to offer, scientific information, depends heavily on the integrity of the scientific process. Scientific information is reliable only if the methods used to derive it and the scientists who interpret it are reliable. Scientific facts do not emerge apart from process.

The scientific process followed in the RF bioeffects field has been reliable for the most part. Researchers have subdivided the main problem—how RF radiation affects living tissue—into subproblems, designed experiments to solve the subproblems, conducted these experiments, and thereby assembled a significant body of information on RF bioeffects. But the research process has not been above reproach. Over the years individual biases and inconsistencies have raised questions about the integrity of the entire field. Some examples follow.

Radar Death

In March 1954 a forty-two-year-old man walked in front of a transmitting radar installation. Within seconds he felt internal warming and quickly moved. Thirty minutes after the initial exposure he experienced "acute abdominal pain and vomited." An hour later he was admitted to a hospital and within six hours had his appendix removed. Recovery from the operation did not proceed normally. Eleven days later the patient died of complications. A possible cause of death according to John McLaughlin, an attending physician, was "tissue destruction . . . from microwave radiation (radar)."[23]

McLaughlin's diagnosis rested on several considerations. During the first operation physicians observed that "the entire parietal and visceral peritoneum were dusky red and the portion of the small bowel that could be seen was beefy in color," indications that the organs in the patient's body cavity could have been "cooked." Moreover the gross appearance and progress of this patient's illness seemed to follow patterns seen in animals killed by excessive exposure to RF radiation. Herman Schwan and other scientists had already suggested that by the time heat is felt internally, "the tolerable level has been exceeded." The other signs observed did not point decisively to any other cause of death. These factors were sufficient to prompt McLaughlin to recommend that these and others factors should be studied in more detail.[24]

More certainly could have been learned from further study; McLaughlin had no information on either power or wavelength when he made his diagnosis, estimates of absorbed power could have been made, and comparisons to animal studies could have been explored. But more information was not gathered. Instead military officials sent McLaughlin's article to the Armed Forces Institute of Pathology for comments. When a review was returned with the conclusion that this was "not an acceptable instance of intestinal damage due to radar," the case was dismissed.[25]

The unanswered questions surrounding this case troubled some researchers. Herman Schwan tried to get more information but was not successful. He explained during the 1967 Senate hearings, "The medical report [by McLaughlin] was the subject of controversy. It was stated by some parts of the medical community that the type of injury suffered indicated something which was quite out of the ordinary. On the other hand, other people stated that he might have had an appendix condition which was severely advanced. . . . I tried to get from the manufacturer information which would provide me with a clue; namely what was the piece of equipment, and so on. I didn't get the information."[26] As far as industry and the military were concerned, the case was closed. It remained so until 1970, when a physician responding in the *Journal of the American Medical Association* to a question about RF bioeffects listed "death" as one consequence of RF exposure and cited McLaughlin's article for support.[27]

The article in the journal raised understandable concern: McLaughlin's radar death case had been disputed, and the connection between RF exposure and death had not been established. To support this contention the report on McLaughlin's article was dug out and sent to Thomas Ely, a physician at Kodak and a member of C95.IV, who used it as the basis for a reply. Quoting the relevant passages from the report, Ely told

Journal readers that McLaughlin's article had been "thoroughly discounted long-ago."[28]

The same message was repeated in more detail three years later by Sol Michaelson. In an article submitted to the Senate Commerce Committee in 1973, Michaelson attempted to lay this report of radar death to rest. He reiterated Ely's contention that the 1958 report "thoroughly discounted this case," noting as well that it "unequivocally found no evidence for implementing microwaves in the patient's death." According to Michaelson, the death was "in fact, a case of acute appendicitis in which evisceration of the wound occurred on the tenth post-operative day leading to profound shock and death." In case anyone doubted this diagnosis, Michaelson went on to argue that "it is extremely doubtful that such damage could occur within this man's abdomen without coexisting skin pathology [burns]." Since there were no reports of burns, presumably the death could not have been caused by RF exposure.[29]

Michaelson's rebuttal effectively crystallized a one-sided assessment of this case that had been in common circulation since the late 1950s. The 1958 report had not "unequivocally found no evidence" to link the death to RF exposure; it had just found the evidence linking the death to RF exposure unconvincing. The leap from unconvincing evidence to no evidence reflects a judgment, not scientific reasoning. The death had not positively been ascribed to complications following appendicitis. This was one diagnosis among many, as the autopsy made clear.[30] The conclusion about a single cause of death again reflected a judgment, not an established fact. The argument about burns was also a judgment, based presumably on Michaelson's interpretation of his own work. It was not based on scientific evidence, and some scientists even doubted that it was true.[31] In sum the expert evidence Michaelson submitted as a scientist was biased by one-sided judgments, all broadly sheltered under the umbrella of scientific expertise.

The extent to which Michaelson was aware of his own biases is difficult to determine. His critics have accused him of deliberately distorting information in this and in other cases in exchange for lucrative research grants and consulting fees from the military-industrial complex. The alleged discovery of this situation has played a major role in fueling the microwave debate. Michaelson has denied such charges, maintaining that his conclusions have the weight of science behind them. Whatever his motivations the fact remains that Michaelson's scientific analysis of the radar death case was not reliable. Insofar as he has been willing to go beyond fact and interject his own interpretations into the scientific process, his credibility and the credibility of those who support him has legitimately been called into question.

This does not mean that the 1954 accident was a case of radar death. The exposure may have been unrelated to the death; it may have been a secondary causal factor, aggravating a preexisting minor case of appendicitis, or it may have been instrumental in the death. Until more information is collected, this case will remain unsolved. To argue otherwise is not consistent with rigorous scientific thinking.

Epidemiology

Lack of critical scientific thinking has also played a major role in shaping attitudes toward epidemiological studies, surveys of the health of human populations. In principle epidemiological studies are deceptively simple, making it difficult for the public to understand why they have been so neglected. Their main objective is to discover whether select populations that have been exposed to some factor, such as RF radiation, are (retrospective) or will be (prospective) as healthy as similar populations that have not been exposed. If such studies are carefully controlled so that there is only one major variable, then differences in health can be linked to that factor or, conversely, the lack of differences can be used to suggest safety.

In practice epidemiological studies are never this straightforward. Working populations are usually exposed to many variables. Their exposure to any one variable, such as RF radiation, is difficult to quantify. It is difficult to identify stable populations that are large enough to permit the detection of subtle effects. Surveying the health of large populations is time-consuming and expensive. The detection of a moderate effect that takes fifteen years to develop could require monitoring the health of thousands of persons for half their lifetimes.

Despite these difficulties, it has long been recognized that the health of populations routinely exposed to higher than normal amounts of RF radiation should be monitored. Such studies were recommended and carried out at the height of World War II. Epidemiological studies were singled out as important at the

1953 navy conference. The need for more epidemiological studies was included in ERMAC's 1971 program, in a 1978 report prepared for the Office of Science and Technology Policy, and in more recent surveys of ongoing research needs.

These calls for more epidemiological studies have not been followed for the most part. In all only about a dozen such studies have been undertaken in the United States over the past forty years.[32] In addition U.S. researchers have been unwilling to accept the epidemiological data collected in the USSR and East European countries in the course of monitoring the health of workers. The fact that workers in these countries who have been exposed to RF radiation often complain of headaches, dizziness, loss of memory, and other asthenic conditions has not been regarded as significant in the United States, primarily because of the uncontrolled manner in which this information has been collected and reported. Philosophically U.S. scientists have not been prepared to believe that alleged RF bioeffects exist until they can be unambiguously tied to RF exposure.

There has been inconsistency in this attitude, however. U.S. researchers have been most demanding when deciding whether effects tentatively identified in epidemiological studies are significant. Typically the research pattern followed has been to keep checking and rechecking until possible causes for concern have been dismissed in some way. Tests conducted by the navy during World War II were ended when a superficial analysis of the data suggested no major effects. A mid-1950 survey of Lockheed workers was repeated when blood abnormalities were found and terminated when the abnormalities supposedly were traced to an experimental error. An ambitious eye screening program begun during the Tri-Service era was ended when it was concluded that the differences that were found—a higher percentage of minor lens defects in exposed workers—were of no clinical significance. A mid-1960s birth defect study conducted in the Baltimore area was rerun when correlations between radar and Down's syndrome emerged; the study was terminated when the correlation disappeared on closer statistical analysis. A similar course of events took place following initial reports of high numbers of congenital abnormalities in the Fort Rucker area, also in the late 1960s. Two ambitious studies undertaken in the late 1970s, one surveying a population of Korean War veterans and the other looking for possible adverse effects among Moscow embassy personnel, were both terminated when the initial results turned up no apparent correlations between exposure and effects.[33]

In all of these studies critical analysis uncovered reasons to question the possible effects discovered, leading to the conclusion that epidemiological studies had not proved that exposure to RF radiation is hazardous; however, critical analysis did not then continue in an effort to determine whether conclusions could be reached about safety. Although the researchers who have conducted the epidemiological studies have consistently pointed out weaknesses in their work that could have allowed effects to slip through undetected, these weaknesses have not been given the same attention as the shortcomings in the methods used to discover effects. The most recent example of this inconsistent use of scientific rigor can be found in a 1976–1978 epidemiological survey of the health of the population exposed to the Moscow signal.

Shortly after the Moscow embassy problem became public in early 1976, State Department officials decided to quell public fears by asking an independent researcher, Dr. Abraham Lillienfeld of Johns Hopkins, to survey the Moscow population for possible health effects. From the start Lillienfeld's study had problems. Personnel lists were difficult to assemble. (The State Department does not publish embassy directories in sensitive areas.) Government agencies such as the CIA and Department of Defense were reluctant to provide information on their personnel. Data on the signal were not made available so those having the highest exposure could not be isolated. Embassy personnel were slow in responding to questionnaires, yet the State Department insisted that a rigorous time schedule be maintained. In all, the Foreign Service Health Status Study (FSHSS), as the Moscow survey came to be called, did not proceed under ideal conditions. Even so data were collected.

The FSHSS data confirmed the State Department's contention that the Moscow signal was not causing immediate health problems. Embassy personnel apparently were not dying any faster or contracting more illnesses then their control counterparts (personnel in other East European embassies). But it became apparent that the study was not sensitive enough to discover anything but immediate major effects. No conclusions could be drawn from the discovery that "the proportion of cancer deaths was higher in female employees" because of the small number of deaths. The difference between exposed and controlled populations was about two deaths. Similarly no long-term conclusions could be drawn because "the group with the highest exposure to microwaves, those who were present at the Moscow embassy during the period from June 1975 to February 1976, had had only a short time for any effects to appear." This study, like so many other epidemiological studies, was not sensitive enough to allow firm conclusions to be drawn about effects or safety.[34]

Such weaknesses might have been corrected by improving the scientific process, following the pattern set in studies reporting effects. This step was contemplated by ERMAC members in mid-1981 and rejected. The population and the exposure information were judged inadequate for pursuing a more detailed study; therefore the data could not be refined, making it intellectually unjustifiable to spend more money on the project.[35] Thus ambiguities would remain. Correlations between RF exposure and health effects had not been proved but neither had they been disproved.

Members of the scientific community have not been consistent in reporting the inconclusiveness of the FSHSS to the public. State Department adviser Herb Pollack has routinely relied on the FSHSS to assure audiences that long-term exposure is not harmful. A paper he presented in 1980, after discussing all of the measurements that failed to turn up significant differences, concluded that "no convincing evidence was discovered that would directly implicate the exposure to microwave radiation experienced by the employees at the Moscow embassy in the causation of any adverse health effects as of the time of this analysis." Pollack did not provide his audience with the same critical analysis of this conclusion that appeared in the report itself.[36]

An even more blatant neglect of qualifying statements can be found in an environmental impact statement (EIS) prepared in the Bainbridge Island uplink case. (The EIS was prepared by scientists and intended to be used by government and the public when making decisions about the proposed uplink facility.) After describing the FSHSS and noting that "researchers found no reliable differences between health status records of the Moscow embassy personnel and the control group," the conclusion was reached that "these findings indicate that there is no danger from continuous exposure to microwave radiation at levels below 15 microwatts/cm^2."[37] This statement not only ignores the qualifying remarks included in the FSHSS but also presents conclusions that were not drawn and could not have been drawn on the basis of the available evidence.

Such one-sided reporting pervades other portions of the discussion of epidemiological studies in the Bainbridge Island EIS. After expressing the usual methodological reservations about USSR and East European research, the authors of the EIS accept without question a recent summary of USSR and East European epidemiological studies delivered in absentia by a Czechoslovakian researcher in 1981: the relevance of the summarized studies is that they are said to "imply that there is no danger to [*sic*] continuous exposure to microwave radiation levels of $10\mu W/cm^2$ and below." This in turn supports the contention that "there are no substantial data from epidemiological studies to suggest that adverse health effects seen in laboratory animals have occurred in human populations."[38]

Why this particular survey has been accepted when other USSR and East European surveys are rejected is not explained. Neither is it explained why the attention of readers is drawn only to studies that found no effects when the summary mentioned as well studies that found effects. The Czechoslovakian researcher presenting the summary, Jana Pazderova-Vejlupkova, had not observed differences in most health tests run on exposed and unexposed radio station employees, but she did find "a reliable difference in [the] variability of the response to the glucose-tolerance test." Another paper she mentioned reported "no difference in the morphological character of lenticular opacities of exposed and unexposed populations of men," but the same paper offered "the view that long-term exposure to microwaves at power densities below cataractogenic levels can accelerate the natural aging process of the lens." The selective use of data illustrated by these examples hardly seems consistent with a final goal set out in the Pazderova-Vejlupkova paper: "My plea is for adherence to the principle of objectivity."[39]

Admittedly the Bainbridge Island EIS is a particularly biased document and could be dismissed if it had as quickly been dismissed by the scientific community as McLaughlin's radar death article. But this has not happened. Critics of the way RF bioeffects research has been conducted have offered comments, but not the broader scientific community that wants the public to accept its judgments about RF bioeffects.[40] While being rigorous in its criteria for accepting effects or hazards, this community has ignored its own ambivalence toward epidemiological studies and the misuse of them by some of its members. The result of this situation has been a further erosion of the credibility of the scientific community, as viewed from the public's perspective.

Microwave Cataracts

Epidemiological studies and data on individual deaths are controversial by nature; neither can be controlled easily, and neither routinely leads to firm conclusions. The RF bioeffects community is not alone in the problems it has had in dealing with these aspects of science. But the bioeffects debate has not been limited to the softer aspects of science. Controversy has arisen over RF bioeffects that are known to exist, such as the RF or, as it is more commonly called, the microwave cataract.

By the mid-1960s, it was commonly felt that the threshold for cataract formation was about 100mW/cm^2, a figure easily explained using standard thermal mechanisms: exposure at 100mW/cm^2 was known to cause heating, and the eye is known to be particularly susceptible to overheating. The microwave cataract came to be understood as a thermal injury that occurred when the eye is exposed to thermal doses of RF radiation.

Scientific opinion was not unanimous on this subject, however. Russell Carpenter discovered that a single 280mW/cm^2 dose administered to rabbits for three minutes did not produce cataracts. The same dose produced cataracts if administered once daily for three or four consecutive days. It did not produce cataracts if the interval between exposures was extended from daily to weekly, leading Carpenter to suggest that single doses that were not cataractogenic (cataract producing) could produce minor effects that could accumulate and lead to cataract formation if sufficient recovery time were not allowed between exposures. Put more simply, his experiments led to the conclusion that cataracts might be a cumulative RF bioeffect. Subsequent failure to find consistent correlations with temperature rise led to the further suggestion that nonthermal causes might be involved as well.[41]

Carpenter's break with the thermal interpretation of microwave cataracts met with opposition. In a review article published in 1972, Michaelson criticized Carpenter's use of the term *cumulative.* As far as Michaelson was concerned, cumulative damage occurred only when each exposure produced irreparable damage. Carpenter's experiments suggested that the damage was not irreparable since the rabbit's eyes apparently were able to return to normal in a few days, thus explaining the absence of injury with the weekly exposure schedule.[42] A year later, in his review paper submitted to the 1973 Senate hearings, Michaelson made this point more strongly: "It is utterly incongruous that an important concept such as cumulative effect of microwaves should be based on such scanty data. . . . Any scientist can readily see how inappropriate it is to use such inadequate data as a basis for any meaningful analysis." As was his custom Michaelson used the lack of convincing evidence as grounds for total rejection: "Since it has not been conclusively shown, the suggestion of cumulative effects of microwave exposure is untenable."[43]

When Michaelson issued this assessment, more was at stake than Carpenter's cautious excursion into unorthodox thinking. By the time of the second round of Senate hearings, Milton Zaret had advanced a microwave cataract theory that not only challenged orthodox thinking but opened the door to wide criticism of prevailing RF bioeffects policy. In the minds of some, Zaret's views went beyond heterodoxy to heresy and had to be treated accordingly.

The broad outline of Zaret's microwave cataract theory is fairly simple, if somewhat vague. In essence he has repeatedly claimed that long-term, low-level exposure to RF radiation can cause cataracts: "Chronic microwave cataract develops slowly over a period measured in years and follows repeated irradiation at nonthermal intensities; it presents clinically as a gradual degradation of the lens capsule, without any evidence of burn, and resembles a delayed radiational effect."[44] The distinguishing feature of a chronic microwave cataract, in comparison to a thermal microwave cataract, is its location in the eye. Thermal microwave cataracts form in the rear of the lens; chronic (athermal) microwave cataracts, according to Zaret, form in the capsule covering the rear of the lens. The posterior lens capsule, Zaret argues, is gradually clouded by continuous RF exposure, leading to the opacification that characterizes a cataract. The opacification may, in Zaret's view, eventually affect the lens proper, but it begins in the capsule. This distinguishing place of origin, along with some knowledge of prior health, allows Zaret to recognize a cataract as a chronic microwave cataract.

The response to Zaret's microwave cataract theory was cool at best, and as long as it was not taken seriously, no one seemed particularly concerned. But as soon as his theory was cited as evidence for low-level effects, it quickly became the target of attack. An attempt by Leo Birenbaum in 1972 to use one of Zaret's eye surveys to question Michaelson's contention that to date there had been little information on injury brought forth the familiar elaboration of the shortcomings of such surveys.[45] Two years later a concerted campaign to dismiss Zaret's microwave cataract theory was mounted following the publication of a case study in which Zaret potentially linked cataracts in a fifty-one-year-old woman to an allegedly faulty microwave oven.[46] The case study aroused particular concern in the industrial sector of the RF bioeffects community, which at the time was grappling with the problem of growing public distrust of RF technology. The biased responses that followed again undermined the credibility of the scientific process.

Zaret's 1974 report was severely criticized by Michaelson, Raytheon spokesman John Osepchuk, Budd Appleton of the air force, Russell Carpenter, and Harvard ophthalmologist David Donaldson for suggesting that exposure at levels below the ANSI standard could cause injury. "As to the amount of radiation," noted Donaldson, "the incident power is well within the safe limits as specified by the ANSI C95 exposure standard."[47] Appleton was so convinced of the safety of the standard that he was not concerned even with faulty ovens: "Many people in the United States have purchased microwave ovens; many thousands of these appliances are now being used in American homes. Probably some of them leak in excess of the standard currently used. . . . There is mounting evidence that the standard could safely be raised, and although there is not much pressure to raise this standard, exposures in excess of it are being viewed with progressively less alarm."[48] These arguments had no bearing on Zaret's arguments since he was in part reporting the case as evidence for reconsidering the validity of the ANSI standard. To reject his arguments on the basis of that standard and not on the basis of scientific evidence represented a classic exercise in circular reasoning.

The manner in which Zaret's critics handled the scientific arguments in the 1974 report was not much better. Rather than trying to understand the case as presented, the facts were distorted and misrepresented in an effort to undermine the connection to RF exposure. Thus, the phrase "her near vision was becoming blurred [in 1961]" was rephrased by Michaelson and Osepchuk as "blurred vision," noting that this symptom appeared five years before the patient had purchased her microwave oven. The point of the critique based on the reworded diagnosis was to suggest that cataracts could have been forming in advance of the purchase of the oven. The point of the original diagnosis, which Michaelson and Osepchuk ignored, was to establish that the woman's eyes were examined and reported to be healthy in 1961 except for the very common condition of nearsightedness.[49]

Michaelson and Osepchuk also faulted Zaret for connecting this case to RF exposure when he was certain that only one of the cataracts was a microwave cataract: "It is incongruous for the author to definitely state that there was a 'microwave cataract' in the left eye, and for the right eye a microwave cataract is presumptive. It seems obvious that, if this were indeed a radiant energy-induced cataract, both eyes without question or neither eye would have had it, unless this particular patient always checked her 'leaky' oven with one eye and not the other."[50] The logic of this critique is correct; the facts are incorrect. If the patient had uniform exposure, then bilateral cataracts would be expected; however, Zaret's diagnosis of the cataract in the right eye as presumptive in no way suggested that it was not a microwave cataract. He simply did not know whether it was or was not because the cataract had been removed by another physician before Zaret examined the patient. Thus he had to presume rather than conclude, a fact Michaelson and Osepchuk ignored in their version of the case.

Similar biased analyses go hand in hand with legitimate reservations throughout the remainder of the replies to Zaret's 1974 report. Ultimately all of Zaret's critics adopted the traditional position: none was sure what the cause of the cataracts was, but each was sure that the oven was innocent. To ask that scientists supposedly interested in RF bioeffects think further about this case, especially in the light of the fact that cause could not be assigned, was considered a heresy: "the cause and effect relationship [Zaret described] is totally unfounded and represents an erroneous and dangerous conclusion." Zaret was not given the benefit of the doubt even though he listed his diagnosis as an "impression" and not a "conclusion." No one was willing to take seriously the cautious assessment of New York ophthalmologist George Merriam, which was misleadingly said to oppose Zaret: "As you can appreciate, it is impossible to say with absolute certainty that the case reported was or was not due to microwave exposure."[51]

This conclusion can basically be applied to the general state of Zaret's microwave cataract theory. Claims are frequently made that other physicians have looked at some of Zaret's patients and been unable to see the so-called capsular and incipient capsular cataracts. Such checks have never been run under controlled conditions, and the results have not been published. The negative results simply circulate as folklore among those who are convinced that Zaret is wrong. The air force did conduct an eye survey in the early 1970s, but the published accounts of these studies, which claim no effects, are too vague to allow them to be used for evidence.[52] Zaret's work has been tested in the courts by lawyers[53] but not in the laboratory by scientists, a fact that compelled the authors of the Bainbridge island EIS to conclude in the midst of their criticism of Zaret's work under journalistic coverage: "There is still no consensus on the validity of Zaret's theory."[54] Such consensus will not be forthcoming until the biases against Zaret are dropped and his work accepted or rejected on the basis of objective, scientific analysis.

Politics of Science

These instances of biases and inconsistencies reflect a deep-seated problem that plays a major role in scientific development: the politics of science. Scientists are not immune to social, economic, and political pressures. Gaining support for research, getting articles published, receiving promotions, getting elected to important offices, being asked to advise on government panels, to testify at hearings, or to consult on legal cases—all of the activities of science other than working in the laboratory and thinking—bring scientists face to face with many different pressures. They must live up to the standards set by peers, meet the expectations of employers, be able to gain national and international respect, and so on. As long as scientists and society take care to keep the larger system that supports science from influencing the way science is conducted, politicization is not a major problem. When care is not taken to keep politics out of science, biases, inconsistencies, and other detrimental consequences can result. Two additional examples from RF bioeffects research illustrate how external pressure can be brought to bear on science.

Long-Term, Low-Level Effects

In September 1978 Bill Guy of the University of Washington signed a contract with the air force agreeing to plan a long-term, low-level RF bioeffects experiment.[55] Despite persistent calls for such studies, none had previously been conducted in the United States, leaving the air force in a difficult situation. Radar installations are sources of long-term, low-level exposure. By the late 1970s the public was more and more demanding that the air force explain how it could be sure that its radar facilities were not hazardous if long-term, low-level studies had not been conducted. In response the air force made the decision to fund such a study and turned to Guy for help.

The expense and uniqueness of the projected study placed special demands on the scientific process. The full project, from planning through experimentation to final reports, was slated to run about six years and cost close to $2 million. Plans called for running over thirty tests on 200 rats (100 exposed and 100 control) from shortly after birth to death.[56] Given the limited resources available for RF bioeffects research, the likelihood of a similar test being run in the near future was remote. Therefore this experiment had to be as rigorous as possible and above criticism if it were to be of any use. To spend this extraordinary amount of money—ten times the amount for surveying the health of the 4000 employees in the Moscow embassy study or $10,000 per rat compared to $100 per State Department employee—on an experiment that had flaws would have been a tragic mistake.

By late 1979 Guy's research team began reporting on the experimental procedures that were to be adopted in the full 200-rat study (phase II). Their plans raised questions. It was well known by this time, as Guy himself had argued, that behavioral measures were the most sensitive indicators of RF bioeffects and yet he made no mention of behavioral measures when he published a plan for the full study in the January issue of the *IEEE Proceedings*. Since this article reportedly described all of the biological end points that would be used in phase II, the conclusion seemed to follow that behavioral measures were being ignored.[57]

This omission troubled independent researcher Allan Frey, who had been studying behavioral and neurological RF bioeffects for over twenty years. To find out more about Guy's study, Frey wrote to an official at the National Telecommunications and Information Administration, Robert Frazier. As executive secretary of ERMAC, Frazier was usually well informed on current research developments. The information he sent in reply did not mitigate Frey's concerns. On June 9, 1980, Frey responded with a detailed critique of the proposed long-term, low-level study, calling into question the motivations of key members of the RF bioeffects community.

Frey objected to more than the absence of behavioral tests; he was concerned about the proposed use of pathogen-free animals. A unique population would not allow extrapolation "to the general population of rats, much less man." He regarded the "minimum stress" argument used to justify the use of pathogen-free animals as "a joke": "Can they really believe that a rat catching a cold, etc., is more stressful than technicians repeatedly sticking needles into the animals to get blood samples?" Frey was not convinced that the frequency being used was appropriate. And he felt that little would be learned about two important measures, nervous system function and immunology. Most important he objected to the fact that the strengths and weaknesses of Guy's study had not been debated by the scientific community: "Is there any wonder some members of the media believe that there are conspiracies afoot? Won't our scientific community's lack of protest be construed as tacit complicity in a conspiracy? What can we as members of the scientific community offer as a defense—the Nuremberg Defense? If so, we'll also (figuratively) hang!"[58]

Such criticisms did not produce open debate; planning for the full 200-rat study remained under the control of Guy's lab and his air force sponsors. But changes were made in the research design as the study progressed. The request for proposal issued by the air force prior to granting the contract for the second experimental phase of the project listed behavior and evaluation of the immune system as two "parameters to be measured."[59] Both measures were included in the full study, which began on September 1, 1980.

Frey still had reservations. The behavioral study included in the full experiment was an open-field test, a straightforward measure of routine activity. Every six weeks the rats were to be placed, one by one, in a cage designed to measure movement. The cage was divided into squares, each with a photoelectric detection system attached. As the rats moved from square to square, they disturbed the photoelectric system. Each monitored disturbance was fed into a computer and recorded as a measure of the rat's activity. Comparisons of the total movements of exposed versus control rats during one three-minute test session every six weeks formed the data that were used for making judgments about possible behavioral effects.[60]

The addition of the open field test raised new concerns in Frey's mind. At an October 1982 conference I organized to discuss the applicability of risk-benefit analysis to the RF bioeffects field, Frey argued that the open field test was one of the least sensitive behavioral measures that could have been chosen, a view shared by another conference participant, Rochelle Medici, who specialized in studying behavioral effects. If this were the case, then the questions about research ethics remained. Was it unreasonable for the public to believe, Frey asked, "that microwave bioeffects research is at present being channeled to look in the wrong place for effects and . . . that the decisions that have led to this state of affairs were not made in the spirit of science?"[61] Was politics entering into RF bioeffects research?

Guy's response to Frey did not resolve the ethical problems raised. In denying that Frey's criticisms had played a role in shaping the final study, Guy stated that he "did not select the endpoints" for his study. "These were selected in the . . . statement of work contained in the Request for Proposals (RFP) disseminated by the Air Force." In other words Guy deflected Frey's criticism by shifting the responsibility for planning from himself to his air force sponsors. In Frey's view this did not absolve Guy from responsibility. "Why," Frey queried in a follow-up comment, "did Guy take on a project which involved the expenditure of approximately $1.5 million of public funds with the known critical tests ruled out by the sponsor . . . ? Is this science?"[62]

Guy did not agree that known critical tests had been omitted from his study. From his perspective "the behavioral endpoint most sensitive to *low level chronic* microwave exposure, as reported by Professor Shandala, who is responsible for the Soviet microwave exposure standard for the general population, was incorporated into the research plan."[63] This claim is debatable. There are significant differences between Guy's work and the experiments of his Soviet colleague.[64] But even if this problem were overlooked, Frey's basic question still remained: Was it appropriate for the agency seeking scientific advice to select the end points that were used in the study?

The formative role the air force played in Guy's study is not an exception. During the Tri-Service era most decisions about the future of RF bioeffects research were made by the military or within contexts overseen by the military. Much the same support structure was retained in the 1960s, adding the State Department and intelligence communities. Even after the injection of public interests in the late 1960s, the majority of support for RF bioeffects research still came from the military. Throughout the 1970s approximately two of every three research dollars spent on RF bioeffects research can be traced to the navy, air force, or army.[65] And with this support inevitably came control.

Blood-Brain Barrier Effects

In 1975 Frey advanced the hypothesis that low-level RF exposure (below $10mW/cm^2$) can disturb the barrier that regulates the exchange of substances between the blood and brain tissue. This blood-brain barrier is critical for keeping the brain's environment stable and operating normally. His hypothesis was based on a deceptively simple experiment in which anesthetized rats were irradiated with low-level RF radiation (CW and pulsed), injected with a fluorescein dye, and sacrificed. The brains of the rats were then sectioned and examined under ultraviolet light for fluorescence. Comparisons of exposed versus control rats turned up significantly more fluorescence in the brain sections of the exposed rats, a result that led Frey to conjecture that low-level RF exposure might cause the blood-brain barrier to leak.[66]

Additional work on this effect was soon carried out by other researchers, and preliminary reports tentatively confirmed Frey's original hypothesis. A George Washington University researcher, Ernest Albert, using horseradish-peroxidase protein tracer instead of fluorescein, found that $10mW/cm^2$ radiation caused increased barrier permeability in both rats and hamsters. Kenneth Oscar and Daryl Hawkins, two army biomedical researchers, used radioisotope techniques and came up with similar results. Thus by mid-1977, the blood-brain barrier effect had tentatively been confirmed through three fairly independent lines of research.[67]

The reported discovery of this effect met with mixed reaction. Basic research scientists, such as Frey and Albert, were interested in the effect as a possible explanation for behavioral effects. Frey was clearly searching for mechanisms to explain behavioral effects when he began his work. However, from a policy standpoint, barrier effects posed problems, particularly if they occurred at exposure levels below 10mW/cm^2. Changes in the barrier, or in related phenomena such as blood flow in the brain, as a result of exposure at a presumed safe level could not be dismissed as trivial.

The process of assessing these consequences and deciding what to do began in military circles. In April 1977 a secret conference was held at the U.S. Naval Academy in Annapolis, Maryland, to discuss "undesirable electromagnetic effects." At this conference a research team at Brooks Air Force Base in Texas headed by James Merritt reported finding "significant leakage of fluorescein into the brain substance" of rats after they were exposed to a 50 millisecond dose of high-level RF radiation. The brief, high-level exposure delivered enough energy to the brain, Merritt contended, "to increase brain tissue temperatures significantly" and cause "behavioral and anatomic changes." Whether these tests were designed to replicate any existing or projected exposure situations was not indicated in the unclassified abstract of Merritt's paper that was made public.[68]

Several months later Merritt reported on another aspect of his work at a special session on the blood-brain barrier effect convened during a symposium on the Biological Effects of Electromagnetic Waves. In addition to his high-level experiment, Merritt had also attempted to replicate the work of Frey and Oscar and Hawkins. In both cases he failed to get similar results. At low-level exposure his control and exposed animals exhibited similar barrier leakage. The only differences Merritt found were ones produced by heating test animals in an oven to raise body temperature or after injecting the animals with a substance (urea) known to produce a barrier effect. These results led him to conclude that Frey had probably seen a thermal effect on the blood-brain barrier produced by exposure levels that were much higher than had been reported.[69]

The conflicting negative and positive reports on low-level barrier effects prompted the navy to convene a third conference to discuss this problem in October 1978. Once again an effort was made to review all prior work for the purpose of making decisions on the future of barrier research. All of the usual participants were present—Frey, Albert, Oscar, Merritt, and others who had conducted related research. The format called for the presentation of papers, discussion, and then a final summary by Don Justesen, a psychologist at a Kansas City Veterans Administration hospital, who was given this important task even though he personally had not done any blood-brain barrier research. A year earlier Justesen had been asked by a committee of the National Council on Radiation Protection, chaired by Guy, to prepare an analysis of the barrier literature to be used for deriving safety standards. He had also inserted in a special issue of *Radio Science,* which he edited, comments on papers given at a panel session on the barrier. Now he was being asked to comment on yet another summary effort, sponsored this time by the navy.[70]

By this time Justesen had seen enough of Merritt's work to have serious reservations. Shortly after the October 1978 navy conference, he subjected Merritt's 1977 symposium paper, which had been published in *Radiation and Environmental Biophysics,* to a careful reanalysis. He concluded that "there is some discrepancy between your [Merritt's] data and your interpretation of same." Where Merritt had found no effects, Justesen reanalyzed and found them. He also assembled evidence for causal explanations that Merritt had passed over. These objections and others were carefully outlined in a three-page letter written on January 30, 1979. At the time Justesen was under a deadline to produce a review on the blood-brain barrier controversy, forcing him to give Merritt a week to reply.[71]

In theory Justesen's critique of Merritt's work left the controversy unresolved. In practice, however, politics entered the debate and attempted to resolve what science could not. Justesen tempered his private criticisms considerably when he took to print. In a comment on the 1977 symposium papers, which finally found their way into print in late June 1979, Justesen reported that Merritt had discovered his own mistakes, whereupon he "reanalyzed his data via a powerful analysis-of-variance technique and . . . found reliable results." The reported reanalysis also led to several points of interest that bore a remarkable similarity to Justesen's own reanalysis sent to Merritt in January 1979. A year later in a review of the blood-brain barrier controversy, Justesen completely ignored the errors in Merritt's only published paper and presented his own (Justesen's) reinterpretations as though they were Merritt's. This was during the same period that Justesen was wondering in private how Merritt was "going to handle the flat negative conclusion in his abstract [of the 1977 paper] in light of the positive findings obvious in his data."[72]

Justesen was not alone in trying to brush aside the shortcomings of Merritt's work. The final report on the navy's 1978 blood-brain barrier workshop summarized Merritt's work in two brief paragraphs. The first described his experiment. The second noted the negative findings and added, "Merritt's data has [sic] since been analyzed by Justesen and the conclusions are being restudied."[73] More recently a review of blood-brain barrier research prepared at the Stanford Research Institute under an air force contract summarized Merritt's article without any mention of its shortcomings. The same work that Justesen privately took apart piece by piece in his 1979 letter to Merritt had quietly become an example of "more objective and refined detection methods" than were found in Frey's experiments.[74]

The objectives of the preferential treatment afforded not only to Merritt's work but to other studies that have cast doubt on Frey's initial experiments are clear. Justesen's reanalysis endeavored to link the obvious barrier effects Merritt found to thermal mechanisms. Presumably once ties to temperature elevation in the brain were established, the hazards issue was resolved since heat stress was not considered an unusual problem. Viewed from a thermal point of view and using Justesen's estimates, the thermal stresses produced were "no greater than those produced in an animal by the stresses associated with swimming, moderate fasting, learning to escape from an annoying stimulus, or being gentled in the hands of a human being."[75] In essence, it was argued, thermal effects raised no special safety problems. There was no reason to worry about abnormal leakage through the blood-brain barrier if that leakage was caused by heating.

The logic that lay behind this position rested on one debatable assumption—that RF heating is similar to other forms of heating. By the late 1970s it was known that RF energy heats tissues selectively and causes hot spots. The thermal consequences of being exposed to an RF field and being gentled in a hand are not the same. Thus the discovery of possible thermal mechanisms to explain the blood-brain barrier effect was irrelevant to the central question being debated—the relevance of the discovery of a barrier effect at low-level exposure. In addition it was not at all certain how short-term experiments related to long-term exposure. Justesen felt that chronic exposure posed no special problems, but he was not certain. As he wrote Frey, "What will happen after continuous exposure over weeks and months to fields that promote BBB/circulatory changes? That is the big question. My head tells me not much will happen in the way of deleterious effects, but my gut tells me one simply can't speculate away an untested thesis."[76]

Such doubts notwithstanding, the blood-brain barrier controversy was soon speculated away using untested theses. Following the 1978 navy workshop the conclusions that supposedly emerged from the presentations were summarized in a final report. The different results reported by the participants left no doubt "that from a scientific point of view, much more research needs to be done to understand the structure and function of the blood-brain barrier and to evaluate the implications that an increase in barrier permeability would evoke." The state of scientific knowledge was correctly assessed as incomplete. The report then went on to draw a second conclusion: "There appears to be no theoretical or experimental evidence that low-level microwaves that do not raise the brain temperature could be expected to affect the integrity of the barrier."[77] This second conclusion was the untested and also misleading one. Despite claims to the contrary Frey's work had not been replicated by other researchers. It was not true, as the final report claimed citing a 1977 publication, that "Oscar and Hawkins were unable to replicate Frey's results."[78] Merritt had not followed Frey's quantification technique; he had used different techniques for measuring fluorescein leakage. And by the time the workshop summary was submitted, Justesen had already critiqued Merritt's work. Moreover the interjection of thermal reasoning into the second conclusion was irrelevant.

Given the state of uncertainty that characterized blood-brain barrier research in 1978, conclusive evidence either for or against the effect could not have been assembled.

The purpose of the second, untested conclusion is more than apparent from a third conclusion that appeared in the original draft of the workshop summary: "Department of Defense funding of research evaluating the effect of microwaves on the blood-brain barrier should be of low priority." This was the justification used to curtail barrier research in the years to come. This open announcement of policy broke with established tradition, however. Most planning in the military is done in closed meetings, and the results are usually not publicized. The break with tradition was quickly recognized by one of those reviewing the conference for the military, who wrote in the margin of the draft version of the workshop summary: "do we really want to say this?"[79] Apparently they did not; the third conclusion was omitted when the final report was submitted in May 1979.

Frey tried to keep the scientific debate alive by continuing his research and at every opportunity questioning the work of others, an increasingly difficult task. When Justesen failed to publicize his doubts widely, Frey brought the information to the attention of the scientific community in a letter written for publication in the Bioelectromagnetics Society's *Newsletter*. The *Newsletter* editor, Tom Rozzell, at first denied publication, arguing that the piece Frey had submitted was too long. Later he rescinded this decision and allowed the publication to go forward. At the same time, as an official at the Office of Naval Research, Rozzell found himself in the position of informing Frey that his research contract with the navy was to be terminated. In brief, while Rozzell was fostering scientific debate through the *Newsletter,* he was also restricting it by helping to control the flow of vital military research funds.[80]

One year later, in October 1981, Frey ran up against another obstacle when the editor of the Bioelectromagnetics Society's *Journal* and navy employee, Elliot Postow, refused to publish a paper investigating the effects of RF radiation on the blood-vitreous-humor barrier in the eye. The justifications Postow used were one equivocal negative review by Justesen (Russell Carpenter had submitted a favorable review) and his own objections to the "tongue lashing" Frey had given Spackman, Preston, and Merritt.[81] The frank and open criticism that had become a characteristic of Frey's publications over the years did not fit with the conventional and conservative style that had come to dominate RF bioeffects research. Frey was not playing the research game according to the rules.

Ramifications

The attempts to limit the circulation of Frey's ideas and other potentially controversial aspects of RF bioeffects research have worked to the disadvantage of both science and policy. In refusing to air its own internal disagreements in public, the establishment that controls RF bioeffects research has misled the public and researchers. When criticized by Frey for not pointing out weaknesses in Merritt's work, two researchers who recently published a survey of the state of BBB research could only reply: "No further details [on the statistical inaccuracies in Merritt's work] were included in the reports we read, and so we left it to the reader to decide the significance of their results, since we could not elaborate." Had Justesen and others been as open in their critique of Merritt's work as they had been in rejecting Frey, this would not have been the case. The same reviewers also claimed that they were not aware of Frey's own criticisms of that work because the latter had not been published in the scientific literature, a shortcoming that was certainly not the result of Frey's unwillingness to debate the issues in print.[82]

Policy deliberations too have been hampered by the politicization of RF bioeffects research and by the biases and inconsistencies discussed earlier. The knowledge that key decisions on such research have been influenced by persons with vested interests in technological expansion raises questions about the legitimacy of the research. Once the objectivity of science is thrown into doubt, its utility in policymaking is destroyed.

The destructive consequences of politicizing science are well known. Charles Süsskind fully understood what could happen when he agued in 1968, "The lesson is that we must evolve methods of managing and coordinating research without inhibiting the investigator's freedom." He and his colleagues knew that "you must have independent research if you expect to get results that can be believed."[83] But despite such warnings politics and science have become intertwined in the microwave debate. The scientific community has allowed social, economic, and political pressures to influence its activities, thereby destroying the credibility of its product. With the undermining of the credibility of science, the second avenue for resolving the microwave debate (government being the first) became impassable. And perhaps as important the glaring inconsistencies that have emerged over time have brought the mass media and the public into the microwave debate.

8

Mass Media and the Public

What must the press do to get the truth out to the public about low-level radiation? Well, we simply have to keep digging; check, double-check, and if possible triple-check every press release, tip, and press briefing; remain always suspicious of authority, and never lose sight of that old Latin admonition, *Illigitimati non carborundum*—don't let the bastards wear you down.—William Hines, *Annals of the New York Academy of Sciences* (1979)

Some scientists and policymakers who have had to defend past policy decisions during the course of the microwave debate argue that the debate itself was started by and has been kept alive by the mass media and their readers.[1] There is some truth to this argument. Had the press not discovered and publicized the GE television set incident, there is no guarantee that Congress would have acted. Press stories about defective microwave ovens played a crucial role in bringing RF technology under the radiation control umbrella. Coverage of the Moscow embassy radiation problem played a key role in stirring up public interest.

It has also been argued that the main motivation for publicizing the microwave debate has been economic; the microwave story sells newspapers. Again there is some truth to this argument. Popular writers have used the microwave story to sell copy. But there are also more fundamental reasons that account for mass media involvement in the debate. As a managing editor of the Eugene, Oregon, *Register-Guard* explained to readers when defending his decision to print a series of stories about a microwave signal discovered in late 1977:

If the possibility exists—and some researchers believe it does—that microwave radiation beamed at us constantly by radar, broadcast transmitters, telephone relays, satellites, power lines, microwave ovens, CB radios, and other sources is harmful, then no newspaper with any backbone at all is going to sit silently by when a strange, seemingly powerful, radio signal is detected by a group of concerned engineers, physicists and technicians whose credentials are valid.

Believing that "an informed public will respond actively and intelligently to community problems and concerns," the *Register-Guard,* along with other major newspapers in the area, took on the task of educating the public about the mysterious signal and its possible impact on public health.[2]

Similar decisions have been made by other editors and producers. Over the past decade the mass media have covered hundreds of stories and generated thousands of articles, reports, and programs on RF bioeffects. Their objective has been to provide information; their justification, to ensure that the public knows.

Public Discovery of the RF Bioeffects Problem

Events of the late 1960s sharply recast the background against which reports on RF bioeffects were judged. When journalists covered the RF bioeffects story in the late 1950s, concern was being raised mostly by the military over the possibility of future problems. With the exception of a few extraordinary cases, such as the one reported radar death, no large-scale current problems seemed to exist. When journalists covered RF bioeffects in the late 1960s, the focus of concern had shifted to current problems. Even if no one had been injured, X-ray–emitting television sets were in homes, and some microwave ovens in operation had been found to be leaking more than the commonly accepted safe levels of RF radiation.

The immediacy of the RF bioeffects problem by the late 1960s is more than apparent in the headlines that brought the defective television sets and leaking microwave ovens to the public's attention. The actions reported were current: "General Electric will modify 90,000 large color television sets against possible x-radiation leaks."[3] The terms *health hazard* and *peril* became common in all reports. Most important there began to develop the feeling that those in charge in industry and government had not acted with complete responsibility, forcing GE, for example, to defend its "failure to warn TV owners of radiation peril."[4]

Mild criticism of government and industrial actions continued into the 1970s, coming to an initial critical point in 1973 when Consumers Union reported to its readers that "microwave ovens . . . leaked radiation at levels we can't be sure are harmless." Uncertainty in this case prompted CU to stamp "not recommended" on its microwave oven report.[5] Although spokesmen from the microwave oven manufacturing community were and have continued to be critical of CU's report, there were good reasons for considering it responsible. Throughout the early 1970s many respected scientists and government officials argued that too little was known about RF bioeffects to reach definite conclusions about safety. CU simply repeated this claim and urged caution: "Until much more evidence is available regarding the safety of low-level microwave radiation, we do not feel we could consider a microwave oven acceptable, unless there is *no* radiation leakage detectable."[6] Given the legal ramifications of an endorsement and the subsequent discovery of hazards, it is not difficult to understand why CU adopted a cautious stand. If its readers wanted to conclude otherwise, they were free to do so, using the information contained in the report.

Such caution was not limited to CU or microwave ovens. Beginning in 1971 *Washington Post* writer Jack Anderson attempted to alert the public to possible RF problems in the military sphere. A March 1971 story—"Spy Plane Problem"—in his nationally syndicated column claimed that the air force's new EC-121 spy planes were "so loaded with electronic gear that the microwave radiation may be causing eye damage." His source for this story was a Public Health Service study of sixty-four crew men who worked aboard the planes.[7] One month later Anderson relayed a stern warning on the possible danger of microwave cataracts given by Milton Zaret at a meeting of the IEEE. Again connections were made to the EC-121 spy planes.[8] Two months later a new dimension was added to these reports when Anderson learned that a military official reportedly had been silenced by the air force for speaking up too strongly in public about the possibility of RF hazards. Although Anderson did not suggest a major cover-up, the article certainly raised suspicions.[9]

Over the next few years Anderson discovered additional suspicious information. In November 1972 he wrote that the navy was subjecting fifty volunteers to potentially dangerous levels of microwave radiation in an effort to learn more about possible effects. The same article reported that Congressman John Moss of California had learned that microwaves were causing eye problems among air-traffic controllers.[10] By March 10, 1973, the pale of suspicion had spread to the last untouched area of the RF establishment, the research community. Senator John Tunney, who had recently sponsored hearings on the radiation problem, was reportedly so dissatisfied with some of the testimony presented that he launched an investigation into "the supposed 'objectivity' of certain microwave scientists who have received high pay as consultants for industry."[11]

Two months later Anderson dug up new information that linked microwaves to the cloak-and-dagger activities of the world of international spying: "American intelligence agencies are perfecting bizarre surveillance devices which make James Bond's gadgets look Victorian." The list of weapons in the new counterintelligence arsenal included laser guns, infrared cameras, and microwave bugging devices. This added source of "electronic smog" threatened to ruin "the eyes of spies, Communist diplomats, and innocent citizens who just happen to be in or near the rooms when the hazardous rays are unleashed."[12]

Anderson next wrote that the Soviet Union had been using such equipment to spy on the U.S. embassy in Moscow since the early 1960s. The RF device they used was said to be a pulsed microwave beam directed at the embassy from a nearby building. Anderson also learned that the CIA and State Department were concerned enough about the problem to have instituted a secret research project—Pandora—to study the biological effects of the microwaves being beamed at the embassy. Although the reason for irradiating the embassy was not known, speculations within government circles supposedly pointed to brainwashing as one possibility. If embassy employees could be affected psychologically by a low-power microwave beam, one well below the U.S. standard, the radiation of the embassy was no trivial matter. Fortunately, according to Anderson's report, Pandora's experiments failed to uncover any biological effects and so were terminated. "As yet," Anderson noted, "there is no conclusive proof that low-level radiation is harmful."[13]

With Anderson's coverage of the Moscow embassy–Project

Pandora story, the major elements of the cover-up-conspiracy theory that would shortly capture the attention of the media had been made public. Industry had been implicated in the manufacture of potentially hazardous electronic products and shown to have mixed loyalties when it came to public protection. Government officials appeared concerned by the RF bioeffects problem but ineffective in dealing with it. Military planners were thought to be aware of potential hazards and unwilling to deal with them openly. Research scientists had been accused of sacrificing scientific objectivity for consulting fees. All of the pieces of a first-rate cover-up story were in place; however, none of these pieces captured the large amounts of public attention needed to make the RF bioeffects problem into one of national or international dimensions. For this, one more critical ingredient was needed, the full public disclosure of the Moscow embassy problem, which occurred in early February 1976, following the State Department's decision to install aluminum screening in the windows of the embassy as a protective measure against possible adverse health effects. The extraordinary step of installing screens in the middle of a Moscow winter made it impossible to keep the signal problem secret any longer.

Public Disclosure of the Moscow Signal

On February 4, 1976, the U.S. ambassador to Moscow, Walter Stoessel, called and then cancelled a special staff briefing to inform embassy employees about the Moscow signal and the measures being taken to protect against possible health hazards. The unusual step of cancelling a special briefing set the rumor mills within the Moscow compound in operation. By late afternoon the next day Stoessel urgently requested permission from his superiors in Washington to go ahead with the briefing, noting that the "questioning and speculation within the embassy as [a] result [of] calling [the] briefing and then cancelling [it] without explanation continue to grow. Some employees are beginning to focus on possible environmental health problems. For example, one officer informed admin[istrative] counselor this afternoon that if [the] situation [is] not explained soon, he would have no alternative to requesting immediate transfer." Stoessel recognized "that briefing in any form raises [the] risk of leaks to [the] press" but saw no alternative.[14] Aluminum screening, staplers, tin snips, measuring tapes, and razor knives had been shipped to Moscow and were already a source of local gossip.

News of the discontent caused by the cancelled briefing prompted Stoessel's superiors in Washington to grant him immediate permission to inform embassy employees about the microwaves. He was told to stress in the briefings the "importance of avoiding press stories at this critical juncture." Employees were to be informed that the problem was on the "way to solution but that speculation about it in the press might undo the solution."[15]

Despite the warnings, which were relayed to embassy employees when the briefings finally were held on February 6, news of the microwave bombardment became public within a few hours of the first briefing sessions. Late that evening Stoessel sent another urgent wire to Washington warning his superiors that "Robert Toth, Moscow correspondent for *Los Angeles Times*, called me at 11:15 P.M. to say that he had 'full story' on embassy briefing conducted today, February 6. He is filing story tonight."[16]

Toth's story appeared on the first page of the Saturday morning edition of the *Los Angeles Times* under the headline: "Radiation hazard seen in Moscow embassy bugging: U.S. ambassador reportedly warns his staff in secret briefing, offers opportunity to transfer." The next day the *Washington Star* carried details on the briefings. By Monday, February 9, news of the Moscow signal was featured in most major U.S. newspapers.

The State Department did little to clarify the Moscow signal problem, even when it became a matter of public information. At a Monday morning briefing a State Department spokesman told the press that he was not able to comment on reports about the embassy situation "because a decision is [*sic*] made that we are not going to comment on them."[17] On Wednesday a briefing revealed that the embassy doctor, Colonel Arnold Johnson, had met with "members of the American community" in Moscow and "assured them that they had not been exposed to any health hazards."[18] The next day Secretary of State Henry Kissinger provided what became the official reason for the silence: the matter was one "of great delicacy which has many ramifications" and therefore could not be dealt with openly.[19]

The lack of official information did not stop stories from circulating in the press about possible health effects being experienced by embassy personnel. The *Boston Globe* reported that Ambassador Stoessel was suffering from a blood ailment that was possibly aggravated, if not caused, by the microwave radiation.[20] Next it was learned (as Jack Anderson had already reported) that the State Department had known about the radiation for fifteen years but had not told the Moscow staff about it. Details on Pandora also started to make their way into the public sector, largely through the articles of Associated Press writer Barton Reppert.[21] In April news surfaced about the planned health survey of all present and past employees in Moscow, presumably for the purpose of demonstrating that no problems existed.[22] A few weeks later information from secret sources revealed that two children of embassy employees had been sent back to the United States for tests relating to blood disorders.[23] Even without detailed information from official briefings, the press was able to uncover and present to the public an account of related events.

As this information slowly trickled out, the State Department tried to retain the overall general impression that the problem was under control. When events emerged that seemed to suggest that the situation was not under control—for example, when it became known in late February that unusual steps had been and were being taken to strengthen the medical staff in Moscow—the State Department simply returned to the general impression to explain what was going on.

Q. Are you prepared today to tell us what those ramifications might be?

A. No, I am not.

Q. Why are you not telling us?

A. Because I am not able to expand on what the Secretary said.

Q. Why not?

A. Because it is a matter of great delicacy.

Q. What is the delicacy?

Q. If nobody is sick, why is it a matter of great delicacy?

A. Because it has many ramifications—has more than one ramification.

Q. Well, aren't you really putting us into a squirrel cage on this issue? Isn't it the intent . . .

A. No, I am not putting you into a squirrel's cage. You may be entering it by yourself, but I don't think I am.

Q. Well, no. We talk about a matter of great delicacy, and then you turn around and talk about ramifications. When we ask you about ramifications, you talk about delicacy. Is this the way to handle the American public?[24]

Apart from the obvious element of obfuscation inherent in the official responses to reporters' questions, the official story proved frustrating to outsiders because it did not make sense. The public was being asked to believe that the overwhelming concern was the health and welfare of the State Department and other personnel in Moscow. Consonant with this concern the State Department "made unilateral efforts to reduce any dangers."[25] But what dangers were there to reduce? According to the official account given to embassy employees, after the pressure of press speculation demanded that more information be made available to the staff in Moscow, the precautionary measures had been taken "as soon as it was realized that the duration and intensity of the new signal turned on in 1975 could be further increased because of the new Soviet installations."[26] In other words the unprecedented steps taken in February 1976 were designed to protect employees from a future situation that might exist, not from a present danger that did exist. But this benign reading of the situation did not square with the facts.

It is true that changes in the Moscow signal beginning as early as 1973 ultimately brought the signal problem to a climax. In January 1973, after about a decade of irradiation by the original signal (called TUMS—technically unidentified Moscow signal), a second signal was picked up. Dubbed MUTS (a variation of TUMS) this second signal was observed only until March 1973 and then disappeared. It reappeared briefly in February 1974, by which time intelligence sources indicated that the Soviets were constructing special facilities on a building across the street from the embassy to house MUTS. This construction project was carefully monitored and pictures of the additions regularly sent back to the United States (figure 6).[27] In May 1975 MUTS reappeared on a regular basis and was joined in October 1975 by a second beam, MUTS-2, located on the top of a building south of the embassy.[28] This sequence of events led the State Department to send a special delegation to

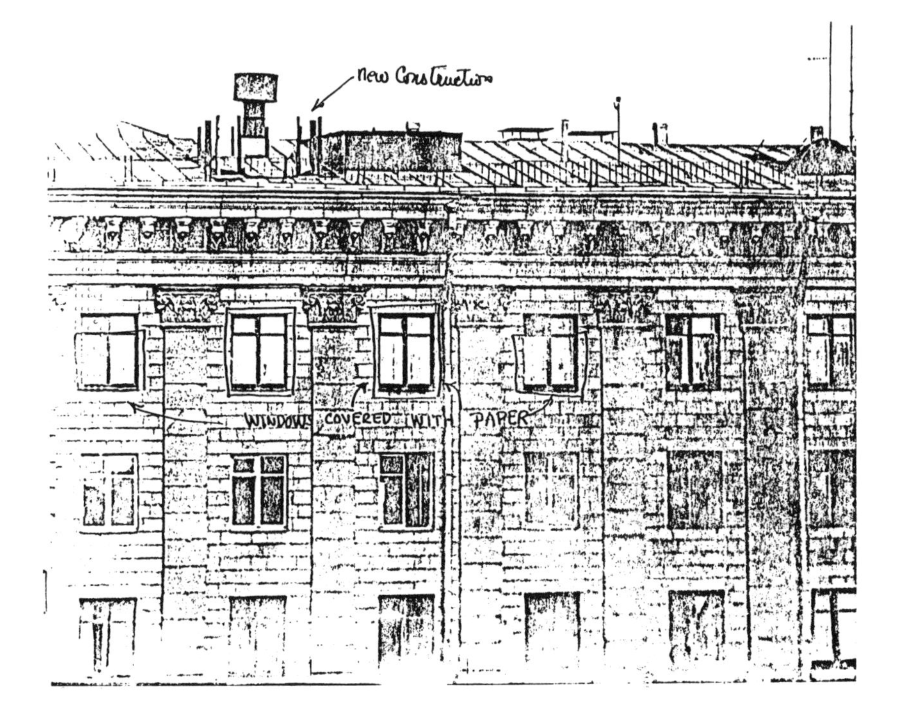

Figure 6
The suspected source of the Moscow signal. From documents released under Freedom of Information requests, USSD MW II, no. 74 (June 19, 1973).

Moscow in early 1976 to negotiate with the Soviet Union in an effort to get MUTS turned off (TUMS was turned off on May 26, 1975).

The State Department's actions unquestionably were motivated by concern over the MUT's characteristics, not its power level. MUTS, even when MUTS-1 and MUTS-2 were operating in tandem, was not appreciably more powerful than TUMS. Both were regularly monitored in the low $\mu W/cm^2$ range—on an average between 2 and 10 $\mu W/cm^2$. But MUTS was more complex than TUMS, and it is this fact that must have been the most troublesome. The State Department, noted one secret cable, was "especially interested in changes" not only in "signal level" but in "composition or operating mode" as well.[29]

The fact that the State Department was more concerned with MUTS's "composition and operating mode" than its strength suggests that health concerns probably were not the main reason for opening negotiations in an effort to get it turned off. Persons such as Herb Pollack and Sam Koslov, who headed the State Department's negotiating team in Moscow and regularly advised the State Department on this matter, did not believe that exposure to very low level RF energy was hazardous. They could not have argued otherwise if they took their own characterizations of the results of Pandora and the Moscow Viral Study seriously. If MUTS presented a new situation, it had to be for very specific and subtle reasons that rested on the unique modulation and composition of this signal.

A belief in special biological effects stemming from specific modulations was not out of the question. Researchers had known since the early 1970s that a properly modulated signal could interfere with normal brain activity.[30] This notion from time to time was picked up and expanded into various mind-control hypotheses, one of which was making its rounds in government circles at the time the Moscow embassy crisis was being discussed by the press. But there is no evidence to suggest that these hypotheses were taken seriously. In fact, the general tenor of opinion seems to have been just the opposite. When congressional researcher Chris Dodge sent one bunch of mind-control papers to Herb Pollack, he appended the note: "Another series for your 'horror' folder."[31] Such an attitude, if it prevailed widely, was hardly the basis for launching a serious protest on health grounds.

But if not for health reasons, why then did MUTS create such a stir? The only other plausible explanation for the actions taken in late 1975 and early 1976 would be a concern over possible advances in Soviet espionage technology. If MUTS was not designed to affect the health of U.S. employees, then its most likely purpose would be to gather intelligence or block intelligence gathering.

Advances in intelligence gathering (or jamming) capability by an opposing power is a delicate matter with many ramifications. It is delicate because a country engaged in the same or similar activities cannot always protest an opponent's actions openly. There are many ramifications because in the world of international spying, every move has many interpretations and many possible countermoves. This being the case, the United States apparently decided to use the health issue as an excuse for protesting MUTS during the official negotiations that got underway in late January 1976. To this end Herb Pollack arrived in Moscow with a copy of a paper on effects of radiation. It was understood that if necessary the "environmental/health agreements" being negotiated between the United States and USSR would be used as a "cover to explain the presence of the delegation."[32]

Protesting MUTS on health grounds could not have been a simple task for the U.S. negotiating team. Its members could have argued that the Soviet Union should not bombard U.S. personnel with radiation levels that approached and occasionally exceeded their most conservative exposure standard ($10\mu W/cm^2$). But the Soviets could easily have countered that the United States had no right protesting a signal approximately 1000 times below the commonly accepted U.S. exposure standard ($10mW/cm^2$) and therefore harmless by U.S. safety criteria.

How the United States negotiated its way around this apparent contradiction has not been made public. Whatever the strategy used, it came close to working. Just as the State Department was about to install the screening in the embassy windows, Ambassador Stoessel received a telegram from Washington telling him not to hold the briefing scheduled for the morning of February 5: "Do not repeat not proceed to brief Embassy or take other contemplated actions for the moment. There may have been a break which we will inform you of by sep[arate] tel[egram] ASAP. The purpose of this message is to emphasize that there be no repeat no briefing until further notice."[33] The order to cancel arrived in time to prevent the briefing from being held but not to stop the announcement of the briefing.

Rumors and speculations spread rapidly. When the story made its way into the press, the State Department had to advance some explanation for the screens and negotiations. The explanation advanced was the same contradictory one used to try to get the signal turned off: concern over health. Rather than opening the delicate issue of intelligence gathering, it was decided to explain the protest action as a conservative move to ensure the future health of embassy employees. This is one conjecture based on the limited documentation made public.

From the State Department's perspective, the conservative approach to the signal problem seemed eminently reasonable; from the employees' perspective it did not. If there were enough concern to conduct secret research projects in the past, to survey employee medical records, to send a medical technician to Moscow to conduct blood tests, to put screens in the embassy windows in winter, and to send a special team to Moscow to demand that MUTS be turned off, then it seemed clear that the State Department and its doctors were not sure and therefore could not say with such confidence that there were no problems. The State Department's deeds and words did not square, at least not from the perspective of those subjected to MUTS. As a result the assurances of safety and concern for the health and welfare of State Department and other personnel in Moscow were not accepted. The press, acting for the public, and the employees themselves wanted evidence, not bland assurances.

In late February the State Department attempted to meet the demand for more evidence by instituting a health survey program. This was not the first such program undertaken. For many years Herb Pollack had routinely monitored the health of embassy employees and reportedly found nothing.[34] But his surveys were not controlled and had not been assembled in a form that could be made public. Hence the State Department had to pick up where it left off in the mid-1960s, beginning with a new series of blood tests. When these tests turned up unexplainable changes in white blood counts—the changes were later attributed to viral contaminants in the drinking water[35]—the decision was made to expand the health survey program by undertaking "an epidemiological study of all employees with Moscow service."[36]

By mid-March the State Department had approved plans for a broad epidemiological study and begun the task of assembling a list of personnel who had served in Moscow. Some thought was given to having the Naval Medical Research and Development Command survey medical records and assemble the needed medical data, but this plan was soon abandoned. It was felt that "having such a survey study undertaken by a non-government agency would sanitize it from obvious questions of bias."[37] Eventually the person saddled with the task of providing an unbiased, rigorous assessment of the hazards and safety issue was Abraham Lillienfeld of Johns Hopkins University.[38]

Announcement of the Johns Hopkins study in July 1976 took some of the pressure off the State Department, directing attention away from present problems and toward the possibility of future solutions. In addition as the give-and-take process of press discovery followed by State Department reply continued, other potential controversies were temporarily set aside. RF radiation was discounted as a cause of altered blood counts in Moscow employees because it did not correlate with actual exposure. Employees who were not regularly exposed to the Moscow signal reportedly experienced the same blood abnormalities as exposed workers. Two special cases of abnormal blood counts experienced by young girls who lived with their parents in Moscow were studied in Washington and eventually attributed to other causes. Grumbling in Congress, mostly by Senator Robert Dole, was countered with personal letters and briefings. Questions raised by the union representing the employees, the American Foreign Service Association, were answered and more information supplied whenever possible. Explanations were given for Pandora and the Moscow Viral Study when they were uncovered and widely reported in the press. With the laying to rest of these and other problems, the State Department was able to dampen some of the speculation.[39]

Had the give-and-take process been allowed to continue without additional pressures, the Moscow embassy problem might have been forgotten by the press and public. But late in 1976 a new blast emerged from the press that threatened to rekindle the entire controversy and spread it even more widely. Science writer Paul Brodeur published two long background articles on the microwave debate in the December issues of the *New Yorker*.[40] These articles picked up on the cover-up-conspiracy theory first tentatively suggested by Jack Anderson in the early 1970s and developed it into a skillfully argued, extensively documented thesis. The view others had cautiously suggested on the basis of interviews and superficial reading, Brodeur now attempted to prove using primary sources from the microwave debate itself. A little more than a year later he ensured that his thesis would not be forgotten by turning the articles into a sensational and controversial book, *The Zapping of America*.[41]

Spreading the Cover-up Story

Brodeur's first reports on the microwave debate in the *New Yorker* had their desired effect. Cover-up was clearly on the minds of members of the Senate Commerce Committee when they opened a third set of hearings on the radiation problem in

June 1977. They asked detailed questions about Project Pandora and the Moscow Viral Study. The contradictions inherent in the pleas for more research dollars versus assurances of safety were openly posed to members of the scientific community. The heads of government agencies were called on to account for past actions and inaction. In the wake of the recently disclosed information, constituents were putting pressure on Congress, and Congress put pressure on its witnesses. Secret files were opened—some of them missing key pieces of evidence—in an effort to learn whether Congress as well as the public had been misled by false assessments of safe levels of exposure.[42]

By summer 1977 news of the electronic smog that was reportedly blanketing the nation was featured on CBS's "60 Minutes." Beginning with the views of critic Milton Zaret, host Mike Wallace spent twenty minutes alerting the public to the possible dangers of microwaves. Viewers were told by military and industrial spokespersons that "we've been living" with RF radiation "for a long time, but we have seen no pattern" of injury occurring. They were also told by critics about the State Department's failure to notify employees about the Moscow embassy problem, about a study of seventy former EC-121 radar men that turned up twelve with cataracts, and about the "billions and billions of dollars" that would be required to bring existing radar emissions "down to a safe standard."[43]

Interest in the RF bioeffects problem grew even more in 1978 when *The Zapping of America* was published. Brodeur's book, subtitled, *Microwaves, Their Deadly Risk, and the Cover-Up,* was packaged to sell. Readers entered the RF bioeffects controversy through this warning on the dust jacket: "Microwave radiation can blind you, alter your behavior, cause genetic damage, even kill you. The risks have been hidden from you by the Pentagon, the State Department, and the electronics industry. With this book the microwave cover-up is ended." If this assessment were correct, then the RF bioeffects story was one that demanded major attention. Newspapers around the country, television and radio stations, magazines—in brief, the mass media in general—accordingly gave the RF bioeffects story the attention they thought it deserved.

Reaction to Brodeur's cover-up thesis was radically divided. *Washington Post* reporter Stephen Rosenfeld called the articles "stunning" and claimed "they'll leave you sick with rage."[44]

Others who also found the cover-up thesis compelling characterized the book as a "literate report on a proliferating threat to life," "heroic investigative reporting," "skillful work," and "an invaluable source book for any group concerned with our environment."[45]

For the persons of this persuasion, *The Zapping of America* rapidly became the source concerned citizens were encouraged to read first. It was looked at as the first reliable and balanced account of a cover-up-conspiracy that unquestionably existed. As *Saturday Review* critic Robert Claiborne commented, "Unlike some other writers on environmental hazards, he hardly ever tries to make rhetorical points by going beyond the facts. Where the evidence is conclusive, he says so; where it isn't, he doesn't pretend otherwise. (As regards the cover-up, no pretense is necessary; the evidence is conclusive beyond a reasonable doubt.)"[46] Brodeur's supporters had little doubt that *The Zapping of America* met the criteria of responsible journalism. It aggressively fostered the public's right to know.

Brodeur's critics, while as adamant in damning the thesis as supporters were in praising it, did not fully vent their views in published reviews. One member of IEEE's Committee on Man and Radiation (COMAR) wrote a stinging but not particularly accurate critique of the *New Yorker* articles, which circulated in manuscript under the committee's name. A copy of this critique made its way to the 1977 Commerce Committee hearings and was offered for publication in the official transcripts; the offer was not accepted.[47] Another copy was sent to *Microwaves* to bring to the attention of the editor, who had just written a favorable review of Brodeur's articles, another point of view. Again the critique was not published.[48] A third copy found its way to the Energy Department, which at the time was funding RF bioeffects research. This fact was brought to the attention of Allan Frey, who had incorrectly been implicated in the review as one of Brodeur's informants. Further the review grossly misrepresented Frey's research. At that time Frey had a bioeffects research contract application pending at the Energy Department so he wrote to IEEE asking that it investigate the ethical conduct of COMAR. IEEE did not intervene, choosing to stick to technical matters and to ignore professional conduct.[49] Edited versions of the critique were sent to the *New Yorker* and IEEE's quasi-popular journal, *Spectrum,* but no publication followed. *Spectrum*'s editors eventually published a review of

Brodeur's work, but not the highly critical one first offered for the public record in June 1977. Instead they waited until the book itself appeared and then printed a more cautious, although still negative, review, written by two scientists, one of whom also happened to be a COMAR member. The review was not, however, published under COMAR's name.[50]

During 1978 other negative reviews appeared, most of them cautious. Reviewers doubted that *cover-up* was the proper term to use for describing past actions, but they did not maintain that all past actions were above criticism. One industrial spokesman attempted to get around the cover-up issue by transferring blame from a presumed willfully negligent corporate-military structure to an unconsciously negligent government bureaucracy: "The examples he [Brodeur] gives in the book of reputed cover-ups can be traced more readily to the general bungling and bureaucracy which we've come to know and love in our government, rather than to any conspiracy."[51] Another reviewer, while praising Brodeur for drawing "attention to some important issues," doubted that the overall tone of the book would lead to constructive public debate in the long run. Objections were raised about the style of writing, the lack of documentation, the one-sided reporting, the obvious attempt to exploit public fears, the use of unfounded innuendos, and the neglect of conflicting data, but no systematic, point-by-point refutation of Brodeur's thesis ever appeared in print.[52]

If Brodeur's critics have been reserved in their response for the most part, his supporters have not. Since December 1977 no part of the mass media has failed to get involved. Newspapers began to run general stories on the RF problem even before *The Zapping of America* appeared. A Knight News wire feature in February 1977 was entitled "Microwave Effects Worry Some Scientists." A similarly titled two-part series by Stephen Lynton appeared in the *Washington Post* in July. The *Philadelphia Inquirer* included a long piece on potential microwave dangers in September 1977, illustrated with a suggestive confrontation between two of the field's leading scientists (figure 7). The following April the *Eugene Register-Guard* ran a six-part abridgement of Brodeur's book as part of its coverage of the mystery signal controversy. The book was said to have "become the touchstone for a growing body of scientists and environmentalists concerned about the effects of 'electronic smog.'"[52]

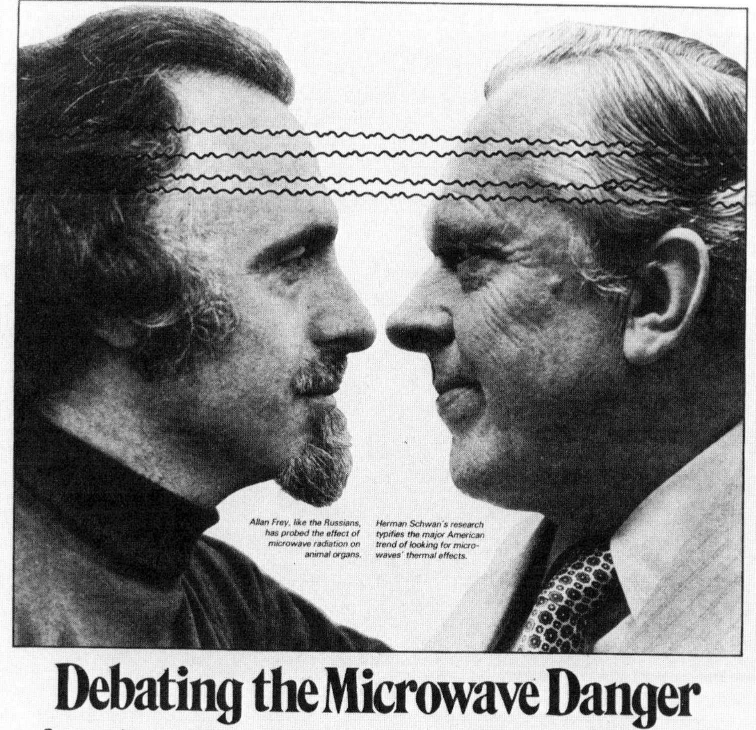

Figure 7
Illustration from the *Philadelphia Inquirer*, September 18, 1977. Reprinted with permission.

Magazine editors began covering the microwave controversy shortly after Brodeur's book appeared. *People Magazine* featured a question-and-answer interview in its January 30, 1978, edition. *New Times* ran an interview-report two months later entitled "The Air Pollution You Can't See." *Newsweek* reported on "The Zap Flap" in July, and *Time* countered in August with "Are Americans Being Zapped? The Microwave Controversy Generates Demands for Action." Subsequent stories, each containing Brodeur's basic arguments filled out with a local example or two, appeared in the *Saturday Review, Science 80, Science for the People, Let's Live, Environment, New York Magazine,* and the *Reader's Digest.*[54]

Radio and television producers have also found the microwave story newsworthy. NBC put together a one-hour docutary, "Microwaves," in 1978. PBS assembled a similar program for radio, called: "Warning, This Program May Be Hazardous to Your Health." A similar documentary produced by BBC was aired on some U.S. television stations. Local controversies, such as the Eugene mystery signal and the Rockaway Township uplink hearings, have been covered by local television and radio stations and occasionally received national coverage.

That the microwave debate has taken off largely as a result of mass media activity cannot be denied. In case after case of public protest over some proposed RF project or problem, the cause of public concern can be traced to information presented by the mass media. Insofar as coverage of the RF bioeffects story in the late 1970s was deeply influenced by *The Zapping of America,* the trail of public concern can ultimately be traced to Paul Brodeur.

The extent to which public involvement in the microwave debate has been changed by the activities of the mass media can be seen in yet another uplink case, this time in Vernon Valley, New Jersey, where public concern over RF bioeffects arose in response to a request by industry (RCA) to build an uplink facility in a residential area. The only difference between the Rockaway and Vernon Valley cases was the fact that in the latter, RCA was seeking permission to build another dish, not the first one. But who had given permission to build the first dish when it was erected in 1974? Essentially no one, because the uplink had been sold as a public utility project that was accepted and approved in the same way as telephone poles and substations. No public hearings were held. No one questioned. No one protested. Before the era of mass media activity in the late 1970s, there was no general public awareness of the potential dangers of RF radiation. And to be sure, in cases where the public did not ask, no one volunteered information on their own initiative.[55]

Insofar as uncontrolled exposure to RF radiation could pose a threat to public health, Brodeur and the mass media have alerted the public to a genuine problem. In so doing they have discharged their first duty as journalists; they have helped the public sector exercise its right to know. But journalism does not stop here. News stories not only alert; they also inform so that the public, once alerted, can "respond actively and intelligently to community problems and concerns."[56] Having aided the right to know, the press has a responsibility to get the facts straight, to bring to the surface divergent points of view, and to provide enough background to allow independent judgment.

This added responsibility has not always been taken seriously by the mass media in the microwave debate. The fine line between reporting and editorializing too often has been blurred as the public has been presented with only the facts that support assumed conclusions. To discern this tendency one has to look no further than to the publication that began general public involvement in the microwave debate, Brodeur's *Zapping of America.*

Brodeur's Thesis

There is some truth to Brodeur's cover-up thesis sketched out in *The Zapping of America.* Project Pandora and the Moscow Viral Study were classified. Code names and cover-up stories were invented to keep from embassy employees the actual reasons for medical tests. The military and CIA have refused researchers, even those conducting the Foreign Service Health Status Study, access to the medical files of employees. U.S. industry has not routinely monitored the health of its workers, even in environments known to approach the ANSI voluntary standard. Government officials have allowed pressure and red tape to stand in the way of instituting a comprehensive RF policy. Scientists have used dual standards and questionable scientific judgment when advising on RF policy. U.S. policymakers in the military, industry, government, and the laboratory have not done all they could to uncover the extent to which RF radiation is hazardous.

If the argument in *The Zapping of America* is taken literally, the cast of characters in the cover-up easily numbers in the hundreds. Entire federal agencies, most of the military, the majority of scientists, and industry are accused of suppressing damaging information. The effects of low-level exposure to RF radiation that have been withheld from the public extend from heart disease and cataracts to birth defects and cancer. Information about such hazards has been withheld to protect the military and its industrial contractors from the economic burdens of lower standards. Such, according to Brodeur, is the magnitude of the conspiracy to cover up and the extent of the opposition the public will face if it tries to change the current state of affairs.

How seriously Brodeur believes in his thesis is difficult to determine. In his book he implicates the FAA in the cover-up because of its unwillingness to recognize "that people working in an aviation environment run a significantly increased risk of developing cataracts," yet when he "flew into New York during a blinding ice storm" he "was glad we had radar in that cockpit." How the risks are to be weighed against the benefits is not made clear.[57] He also continued to build and subsequently occupied a house on Cape Cod after he had discovered that it was being "zapped" by a nearby radar unit with "a power level about a hundred times greater than the power level for which the State Department told its Moscow employees it was installing aluminum window screens for their protection."[58] Although he protested the radiation, he seems not to have regarded it as hazardous enough to warrant moving.

Brodeur employs ambiguity and vagueness as tools to create the sensational cover-up story that has been used to popularize his book. By confusing chronology, taking statements out of context, ignoring evidence or presenting it in negative ways, relying primarily on sources that agree with his point of view, and many other techniques, he is able to craft a history of the development of the microwave debate that suits his purpose and that supports his conclusions. Whether it also mirrors reality is another matter.

In discussing the Moscow embassy crisis, Brodeur interweaves two conflicting scenarios in an effort to raise maximum concern. One scenario argues that the signal was beamed at the embassy at much higher levels (milliwatt instead of microwatt) than were ever admitted. This scenario implicates the State Department in outright deception and adds credence to the hazards charge since it places exposure in the embassy near the presumed thermal threshold of $10mW/cm^2$. His other scenario correlates every malady experienced by the embassy population with low-level exposure in an effort to prove that there are low-level hazards. Which scenario makes its appearance in the chapters devoted to the Moscow embassy crisis depends mostly on the point being made at the time. Brodeur is more than willing to assume that the signal was or was not beamed at low levels if in so doing it helps prove that known hazards were being covered up.

The high-dose scenario is based on two pieces of information: the fact that Pandora experiments were conducted at $4mW/cm^2$ and Milton Zaret's contention that he was told in 1965 that the signal was being beamed "at power densities that never exceeded $4mW/cm^2$." The importance of this information is that it conflicts with public statements made by the State Department after 1976. The later statements always placed the signal's level in the low microwatt range. The conflict between the two reports on levels leads Brodeur to conclude that "either Dr. Zaret's memory is faulty, or the State Department is lying." To help readers decide which alternative is correct, Brodeur reminds them that the State Department had already lied about the Moscow Viral Study.[59]

Had the State Department checked its records carefully in 1976, it would have discovered that early reports (1965–1967) placed the signal at milliwatt levels. Zaret probably was not lying when he said that this information had been passed along to him in 1965. In addition Pandora experiments were conducted at $4mW/cm^2$. Had Brodeur dug more deeply into the Pandora files made public in 1977 or into the State Department files released to Barton Reppert, which he used in other contexts, or checked with Pandora researchers, he would also have discovered that the early measurements were known to have been in error as early as 1967 and that testing was conducted at $8\mu W/cm^2$ and $100\mu W/cm^2$, as well as at $4\ mW/cm^2$. He also could have found that Pandora director Richard Cesaro, who believed that effects were being seen at $4\ mW/cm^2$, reported to his colleagues on September 27, 1967, that at the lower levels "modulated microwave radiation did not cause the primate to degrade in conducting his work tasks."[60]

Rather than evaluating critically the evidence for and against the high-level exposure scenario, Brodeur simply accepted it as true whenever it suited his purpose, as it did when he tried to connect four cases of appendicitis experienced by the embassy staff in 1976 with the signal. Could these cases have been caused by RF exposure? In response to this question, Brodeur reasoned, "It is known . . . that high levels of microwave radiation can have serious effects upon the gastrointestinal tract and upon the organs of the hollow viscera, of which the appendix is one. It is also known that General Wold and his family, as well as the Political Counselor and his wife [the persons afflicted], lived in apartments facing Tchaikovsky Street [facing the signal]."[61] In other words, for the present case he assumed that the signal produced the same high-level exposure that is known to produce internal damage, which presumably is to be understood as thermal damage. He needed the high-dose scenario to link appendicitis to the signal.

Brodeur also had another concern: the effects of chronic, low-dose exposure, which he believed were also evidenced by the Moscow embassy experience. In this case he rested his argument on some hearsay evidence stemming from the Moscow Viral Study.

The first substantial news of the Moscow Viral Study to reach the public came through a series of articles written by Associated Press reporter Barton Reppert. On learning of the study Reppert had contacted the principal investigator at George Washington University, Cecil Jacobson, and was told that the results had been inconclusive. Not satisfied with this generalization, Reppert met with a resident physician in the George Washington Medical School, Thomas Gresinger, who had talked with Jacobson about the results. Gresinger never worked on the viral study but became curious about it when he heard some technicians talking about undisclosed tests being given to embassy employees. When he broached the subject of these tests to Jacobson, Gresinger was reportedly told that they had uncovered chromosome breaks "in significant numbers." This second-hand report (third-hand by the time Brodeur learned of it) forms a key link to an alleged major low-level problem, the cancer connection.[62]

This link is strengthened three chapters later when it is raised in the context of alleged statements by Dr. E. Cuyler Hammond of the American Cancer Society dismissing as "poppycock Buck Rogers stuff" the suggestion that RF radiation might cause cancer. Brodeur uses the evidence from the Moscow study to rebuff Hammond: "What would he say to the fact that when the State Department secretly tested young women from the Moscow embassy for genetic damage during the late 1960s, *it found evidence that such damage had occurred?*"[63] The third-hand report has now become "evidence" for the cancer connection.

One chapter later Brodeur uncovers the smoking gun of the

genetic cover-up in Pollack's report to a Pandora meeting that four Moscow employees "reportedly showed serious chromosomal abnormalities." The fact that Pollack also discussed the inconclusiveness of this report is not mentioned by Brodeur.[64] Nor does he discuss who could have known what, when. It is of no consequence to him that Jacobson did not have the exposure data and therefore could not have made links to RF radiation if and when he did discuss results with Gresinger. Pieces of evidence are added to prove the cancer connection without any regard to their real significance.

One additional link used to forge the cancer connection is a comment made by Zbigniew Brzezinski in a conversation with *Chicago Daily News* columnist Keyes Beech. In May 1977 Beech reported that Brzezinski had told him in March 1976 that the "cancer rate among Americans in the Moscow embassy was the highest in the world."[65] Neither Beech nor Brodeur wondered how Brzezinski came on this information before any systematic epidemiological surveys of the Moscow population had been conducted. For Brodeur Brzezinski's statement simply provided another piece of evidence supporting the cover-up thesis, one that he used whenever it suited him. Thus, thirteen chapters after quoting the Beech story, Brzezinski's authority was again invoked to add strength to a now greatly expanded list of low-level hazards: "sixteen breast cancers, several known cases of leukemia, and a high incidence of blood disorders among people living and working at the American embassy in Moscow, who, according to Zbigniew Brzezinski, the national security advisor to President Carter, are suffering the highest rate of cancer in the world."[66] All of this evidence is now set within the context of the low-level scenario.

Which scenario is correct? For Brodeur this question was of little consequence; all that was important to him was substantiating his cover-up thesis. As far as Brodeur was concerned, the State Department was concealing something; as long as this is believed, the details will fall into place. The same holds true all too often for other evidence. Much of it Brodeur lifted out of context in his drive to construct the case for a broad, self-protecting conspiracy to withhold from the public the truth about RF bioeffects.

Brodeur used quotations out of context to spread the cover-up to BRH and one of its more publicly minded employees, Joe Elder. In 1973 Elder testified on BRH's newly adopted microwave oven standard before Senate hearings on the 1968 Radiation Control Act. He argued that the new standard, which allowed ovens in operation to leak 5mW/cm^2 at 5 centimeters, was a good standard: "we believe we have erred on the side of safety."[67] After reporting Elder's 1973 testimony, Brodeur informed his readers "that less than a year before, as a member of the American National Standards Institute's subcommittee on radio-frequency hazards, he had recommended that the microwave-protection guide be lowered to 1mW/cm^2."[68] The apparent significance of this statement is that it demonstrates that Elder knew that 5mW/cm^2 was not safe when he testified in 1973, thereby implicating him and BRH in the cover-up. If Elder's statements are read in context, however, they do not support this contention.

In testifying before the Tunney committee in 1973, Elder carefully explained that he was talking about "a product standard," not an "exposure standard." Arguing that an oven that leaks no more than 5mW/cm^2 at 5 centimeters is safe is in no way equivalent to arguing that general population exposure at 5mW/cm^2 is safe. Since people do not generally sit 5 centimeters away from an operating microwave oven for long periods of time, an oven that is allowed to leak 5mW/cm^2 at this distance will not expose general populations to continuous 5mW/cm^2 RF radiation. A year earlier when Elder urged ANSI to lower its standard to 1mW/cm^2, he was proposing a general population standard, not a product standard. He felt that continuous exposure under unknown conditions should be limited to 1mW/cm^2. Under known conditions, even in 1973, he was willing to allow exposures as much as ten times higher, or 10mW/cm^2. Elder's position did not change between 1972 and 1973. He was not holding back information when he testified. His remarks can be seen as evidence of a cover-up only if they are taken out of context and distorted.[69]

Brodeur also used one-sided reporting, ignoring or quoting only briefly persons who disagreed with or had been implicated in the cover-up. Persons who agree with his point of view are quoted at length. This is especially true for one of Brodeur's primary informants, Milton Zaret. Brodeur accepts Zaret's word without question and expects his readers to do the same. But should Zaret's word be accepted without question? A careful look at his involvement in RF bioeffects debates suggests that this is perhaps not the case.

Zaret began his long association with the RF bioeffects field during the Tri-Service era as a highly regarded expert on eye problems. His principal and most controversial contributions have been related to microwave cataracts. As one of the few researchers who had studied low-level effects at the time that the Moscow signal was discovered, he was consulted on Pandora, beginning in 1965. In this capacity he not only advised but experimented. In the late 1960s under Pandora sponsorship, he attempted to replicate USSR heart studies and investigated low-level effects using a specially designed test facility on an island in Hawaii. He also served for a time as a member of ANSI C95.IV. Throughout most of this period his credibility as an RF expert remained high.

Zaret's relationship with his colleagues in the RF bioeffects field began to deteriorate in the early 1970s. This was due in part to his controversial microwave cataract theory, which was not accepted by most eye experts. But more important, in 1970 the navy terminated his Hawaiian island experiment under circumstances that questioned his credibility as a researcher. Zaret bitterly resented this move and saw himself as the victim of a military conspiracy to cover up damaging information about low-level RF bioeffects. Soon after he began talking with Jack Anderson, resigned from C95.IV when his advice on lowering the standard was not heeded, and sought redress by testifying on his alleged shabby treatment and on the dangers of low-level effects at government hearings and in court.

Zaret's version of the Hawaiian island episode fits conveniently into the cover-up thesis. He has consistently maintained that his work was terminated by a stacked panel of military and industrial experts who were well known for their closed mind on low-level effects. This panel acted to prevent his promising work on low-level effects from going forward. The main villain in this plot was the navy s Paul Tyler, who, after taking a close look at work being done set the wheels in motion to end Zaret's funding. Thus Zaret became a victim of the cover-up. From this time on the gulf between Zaret and the official RF bioeffects community began to widen.

If this were the full extent of the Hawaiian island story, one might well concur with Brodeur's cover-up thesis and use Zaret as a key witness, but there is evidence on the other side as well. Tyler did not end his review of the Hawaiian island facility with his trip there in 1970. After his visit he sent a review team to the island facility to conduct a full-scale investigation. The report submitted by this team suggests that there was another side to Zaret's scientific credibility.

The Hawaiian experiment was designed to test the response of a colony of monkeys to chronic low-level RF radiation. An RF source was put in a shed on a high point in the center of the island, and the monkeys were exposed in an opening a few meters away. The major problem in the design of the experiment, as judged by the review team, was that the treatment the animals received made it impossible to determine whether RF radiation or other factors caused ill effects that were observed.

During exposure the animals were placed on metal perches in the sun. They were kept on these perches by neck chains that in some instances were too short to allow the animals to reach the ground. Water was not routinely provided. These conditions reportedly led to the death of three monkeys by strangulation, out of a total of five deaths reported. "Leaving animals on their perches overnight" was said to have been "contributory to some of the deaths." Responsibility for this situation was layed not only on Zaret but on the Navy, which was said to have exercised "inept and casual management."[70]

When the unprofessional nature of the experiment was supplemented by other shortcomings in his scientific work, Zaret's credibility as a general RF expert began to be questioned. His attempts to duplicate the Soviet heart-rate experiments had serious methodological flaws, even though they discovered no effects.[71] In 1973 he attempted to tie high heart attack rates in a small Finnish town with Soviet radar just across the border. Jack Anderson picked up this story and Brodeur reported it. Heart attack rates in the area subsequently have been lowered by reducing the amount of fat in diets and smoking.[72]

Brodeur makes very little of this other side of Zaret's credentials. He is more than willing to find fault with those who deny the existence of a cover-up but not with those who support it. He readily points out conflicts of interest when these conflicts explain why the military, for example, has allegedly engaged in a cover-up. He fails to mention that critics of the military can have conflicts of interest of their own. Zaret, for one, receives a substantial fee for consulting and testifying in microwave cataract cases. If it is important to be reminded by Brodeur whenever Michaelson is mentioned that the latter has frequently been retained by industry, it is equally as important to be reminded that Zaret in part makes a living by discovering, treating, and testifying about microwave cataracts.

If a cover-up is the knowing suppression of information, then it can be argued that the United States has been "zapped" by *The Zapping of America.* Brodeur's one-sided, biased coverage of the RF bioeffects debate does withhold information from the public that is crucial for making informed judgments about RF bioeffects policy. In this regard he is no more responsible or objective than the persons he criticizes.

Public Education

It probably is best not to demand too much objectivity from Brodeur. The goal of *The Zapping of America* is clearly as much to persuade and spur to action as to inform. Brodeur took time after publishing this book to use a grant from the Alice Patterson Foundation to study its effect. His book is not a piece of journalism and should not be judged as such; it is a polemical work. Even those who have used it for their own purposes to protest RF policy are aware of this fact. But awareness of the weaknesses of Brodeur's work has not prompted mass media writers to exercise their own critical judgment as much as they should when reporting on the microwave debate.

The education the public has received from the mass media on the microwave controversy has been far from satisfactory. This has been true in all too many cases because the press has simply accepted Brodeur's cover-up thesis at face value, added a few facts to bring it up to date, and sold this package as the latest word on the microwave debate. During the mystery signal crisis in Oregon, readers of the *Eugene Register-Guard* received the bulk of their education on RF bioeffects through serialized installments from *The Zapping of America.* Much less space was devoted to the detailed background information needed to evaluate Brodeur. Few reporters have checked and rechecked their sources sufficiently to make sure that the truth is getting out to the public.

The lack of critical reporting has allowed some of Brodeur's illusions to become popular images. By the time the microwave story made its way into the *Reader's Digest* in 1980, speculations about links to cancer had become fact: "former embassy personnel exhibit a higher rate of cancer than the American average."[73] (The Johns Hopkins study concluded that it was too early to draw conclusions about cancer.) One year earlier Susan Schiefelbein confidently told *Saturday Review* readers that three ambassadors to Moscow had contracted cancer "after serving at that post."[74] How she knew that none had had cancer before serving in Moscow is not reported.

A significant percentage of this misinformation is the result of sloppy thinking. Time and again those reporting the news have not taken the time to consider whether their arguments make sense. In 1978 *Saturday Review* writer Robert Claiborne told his readers that "Brodeur evaluates his evidence . . . in an exemplary manner." One paragraph later readers are informed that *The Zapping of America* is "crammed with great lumps of undigested and indigestible scientific and bureaucratic jargon" and that "at times, the reader finds himself drowning in acronymia alphabet soup and other trivia."

The distinction that supposedly lies behind these two contrasting evaluations is the distinction between the author's presentation and his evaluation of evidence. Brodeur apparently presents his evidence abysmally but evaluates it brilliantly, if Claiborne's argument is to be believed. But can "indigestible" evidence be "conclusive beyond a reasonable doubt"? Claiborne apparently has not thought deeply about this. He has allowed his critical powers to be lulled to sleep by a belief that "after the Pentagon Papers and Watergate, it is hardly news that government officials lie." He has not even thought about the significance of the fact that Brodeur "tells us rather less than we need to know about microwaves themselves."[75]

The absence of critical thinking has unfortunate consequences. News stories do not always separate fact from opinions. Reporting and editorializing are frequently mixed without any indication of where one begins and the other leaves off. Evidence is not always carefully evaluated for its significance and possible biases. Chronology and context are not always taken into account. In many important and seemingly unimportant ways, the end products that roll off the presses or are broadcast on the nation's airways are short on educational content.

The ultimate responsibility for this situation rests more with the editors, managers, and owners of the mass media than with individual reporters and writers. Before a journalist writes a story, a host of small decisions have been made that will determine how well that story can be covered. Space is allocated, audiences are targeted, the amount of time spent collecting background is set. Decisions are made in advance about the level of writing that is appropriate and the sophistication of the reading public.

Through this predecision process, the mass media in America have gradually lost their intellectual vigor. Reporters are not given enough space and time to attempt to educate their readers. Events predominate in reporting, particularly events that are "newsworthy," in the sense of being provocative, controversial, of local interest, and so on. Pieces that pretend to have some intellectual rigor are discouraged. It is assumed that the U.S. reading public cannot be educated beyond some low level—that Americans have neither the time nor the inclination to learn. Rather than accepting difficult problems such as the microwave debate as challenges to the information-education process, editorial and managerial decisions have inclined in the opposite direction. And the larger the circulation and more influential the publication, the more these tendencies have predominated, with the so-called national news magazines providing by far the most superficial and least intellectually challenging coverage of the microwave debate.

Responsible education is a difficult job. There are no short cuts. It is impossible in the space of *Newsweek*'s or *Time*'s science and medicine columns to cover even a fraction of the information needed to think critically about uplink towers, smoking, car accidents, or acid rain. The public must be able to distinguish uplink, down-link, and relay facilities; it must be able to understand how scientists extrapolate from animal experiments and epidemiological studies to standards. Hazards must be distinguished from effects; chronic, low-level exposure from short-term, high-level exposure. The mass media need to accept and rise to the challenge of education or give up the pretense of providing the public with more than a daily chronicle of events.

Robert Claiborne was correct when he criticized Brodeur for telling "us rather less than we need to know about microwaves themselves." The same could be said for mass media coverage of the microwave debate in general; it has supplied the public with rather less than it needs to know. Even so the press will continue to play a vital role in the process of educating the public. The mass media may not provide a complete or unbiased point of view but at least they provide another point of view, one essential for balancing and weighing the equally opinionated information that issues from military, government, scientific, and industrial circles. Democracy would be better served if this balancing force placed a great deal more emphasis on rigor, critical thinking, and elevating the level of intellectual understanding; if the mass media took their responsibility as educators as seriously as they take their right to inform. Until this happens, the public, like the press, must "keep digging; check, double-check, and if possible triple-check . . .; remain always suspicious of authority; and never lose sight of that old Latin admonition, *Illigitimati non carborundum*—don't let the bastards wear you down."

9

Hearings and Litigation: The Last Resort

I never gave up. They wanted me to say there was no connection between Sam's illness and microwaves, but that just wasn't true.—Nettie Yannon, April 1982

The microwave debate has intersected with the law at three distinct points: challenges to the law itself, zoning disputes, and personal injury claims. Challenges to the law have been the least utilized because there have been few laws to challenge. Zoning disputes have played a major role in questioning the safety of proposed RF installations, such as satellite uplinks and radar facilities. Personal injury claims have centered on past exposure situations alleged to have caused health problems. Since the mid-1970s RF-related legal actions have been about equally divided between zoning disputes and personal injury claims.[1]

Problem resolution through hearings and litigation has the advantage of guaranteeing action. Once a legal proceeding is begun, decisions are usually reached. The same cannot be said for scientific, political, or journalistic processes, each of which has ways for avoiding decision making. It does not necessarily follow, however, that decisions reached in legal settings have helped resolve the RF bioeffects debate.

Litigation and Policy

When legislation was first proposed to deal with the perceived radiation crisis of the late 1960s, opinion was not unanimous on such a need; however, since the legislation that was adopted passed the task of policymaking to the executive and did not itself establish policy, legal challenges did not follow. Similarly opinion was also divided on the appropriateness of the executive's first policy decision on RF exposure—the microwave oven standard—but again legal action did not follow. The compromises reached in drawing up this standard seem to have satisfied all interested parties. The same cannot be said for the next and only other decision on standards made by the executive in the 1970s, OSHA's 1971 decision to adopt an interim work place standard.[2]

The validity of OSHA's interim RF standard came into question in March 1975 when it was used to cite the Swimline Corporation in Commack, New York, for a safety violation. OSHA officials had inspected Swimline's facilities in February 1975 and found that one of the RF-sealer operators was being exposed to more than $10 mW/cm^2$ of RF radiation. Swimline contested the citation, thereby sending the matter to OSHA's Review Commission for a hearing and determination. The administrative law judge who heard the case, James P. O'Connell, quickly settled on one crucial issue. He ruled that the standard adopted by OSHA on an interim basis, C95.1, was not an appropriate occupational standard, as defined by law. After setting out legal and grammatical definitions for *should* and *shall*, O'Connell ruled, "I am firmly convinced that the use of the word 'should' in the regulation at issue herein can only mean that the standard [C95.1] was meant to be advisory and not mandatory. Accordingly, respondent [Swimline Corp.] cannot be held in violation of a regulation which is only advisory in form." His findings of fact therefore dismissed the citation and notified OSHA that its adopted standard "is not an occupational safety and health standard as defined by . . . the [Occupational Health and Safety] Act." The law had been used successfully to challenge a policy decision.[3]

Whether similar actions will be taken in the future depends on the extent to which the policy initiatives allowed under the law are used and on the extent to which government is responsive to the advice it gets when formulating policy. Agencies are required to seek advice as part of policymaking, but they are not required to listen to that advice. The FCC was required to seek public comment when it was contemplating setting standards to regulate the RF facilities under its control. It would have had difficulty accommodating all of the advice it received since it was informed that $10 mW/cm^2$ was and was not an appropriate standard.[4]

It also remains to be seen whether other policy decisions relating to RF bioeffects will be challenged. The Rockaway Township Zoning Board of Adjustment identified an obvious target for litigation when they had their lawyer direct a letter to ANSI questioning the basis of C95.1.[5] ANSI's RF standard has played a major role in shaping RF bioeffects policy. Whether ANSI is as a result in part responsible for some of the decisions that have been made using C95.1 could be tested in court.

It might be possible to challenge policy decisions made when funding RF bioeffects research. If these decisions can be shown to have built-in biases and if the results of funded experiments are used to formulate public policy, then the public could intervene and request funding for impartial research. Similarly editorial decisions made by journal editors could be challenged if they had an influence on policy, especially if public funds were used for publication purposes. In sum the uses to which law could be put for challenging RF bioeffects policy decisions are many. The extent to which it is used will probably depend on how successfully legal actions discussed in the rest of this chapter can be resolved.

Zoning Disputes

The unsuccessful attempt of Home Box Office to get permission to build an uplink facility on a rural tract of land in Rockaway Township, New Jersey, is a prime example of a zoning case and of the way RF bioeffects can be drawn into the discussion. When HBO officials discovered that they could not get a quick, favorable decision in Rockaway, they looked elsewhere for a site. They reexamined the target zone around New York City and selected new sites in areas previously said to be unacceptable. The site that presented the least difficulties was a 12.8 acre tract of land in Hauppauge, Long Island. After minor delays and a brief display of public disapproval, the formal go-ahead to build was given on February 11, 1982, a little more than a year after the Rockaway Township application was withdrawn.

Several factors account for the relative ease with which the Long Island plans were approved. This time HBO officials selected an industrial park for their facility, avoiding the sensitive process of applying for a land use variance. They diffused public protest by having a public hearing delayed for two months, during which time a major bone of contention—a 300 foot relay tower—was removed from the proposal. The two-month delay was also used to prepare an environmental impact statement. But perhaps most important was the fact that the Hauppauge zoning board was favorable to development. The board listened to the arguments, accepted the judgments of the experts, and voted to approve. The debate over safety, which had dragged on for months in Rockaway Township, was terminated after a single open meeting and a few weeks of closed-door deliberations.

The Long Island Board of Zoning Appeals made no direct pronouncements on safety in the variance granted to HBO. Legally they had no obligation to do so. The only exemption HBO needed was a height variance; the 45 foot dishes exceeded the 16 foot height maximum allowed in the industrial park and therefore needed special approval. The safety issue did not have to be discussed if there were no reason to question the facility's safety. In granting the height variance the board found no reason to question the safety of the proposed facility.[6]

The process did not proceed as smoothly for the RCA Corporation 3000 miles away on Bainbridge Island. The situation there more nearly mirrored the Rockaway Township case. RCA was planning to build in a rural, residential area, not in an industrial park. Citizen opposition was strong, well organized, and sensitive to environmental issues. The legal system that applied ensured that the RF bioeffects issue would be considered. In sum all of the obstacles that had prevented the Rockaway Township application from going forward were present on Bainbridge Island. The only major difference was that on the West Coast the authority to decide lay at the county level; in Rockaway the affected township made the zoning decision. The county system had the potential for weakening local public input when it was pitted against regional plans for development and growth.

The legal system in Kitsap County, in which Bainbridge Island lies, called for a multistage review. Deliberations began officially on July 28, 1981, when RCA applied to the Kitsap County Department of Community Development for an unclassified public use permit. This application had to be reviewed and passed by the County Planning Department, reviewed by a county hearing examiner, and accepted by the county commissioners before construction could begin. Prior to the County Planning Department's accepting or rejecting the application, advice had to be solicited from the Bainbridge Island Planning Advisory Council (BIPAC) and checks made to ensure that the proposal conformed to the State Environmental Protection Act. The latter process was overseen by the county environmental coordinator, Rick Kimball, who also oversaw the application process.

Local opposition to the uplink project emerged prior to RCA's formal application for a land use variance. Under the leadership of Jerry Hellmuth, the director of a school for children with learning disabilities located near the proposed site, the Bucklin Hill Neighborhood Association (BHNA) started meeting in early July to plan its opposition strategy. When RCA officials arrived at a BIPAC meeting on July 15, 1981 to provide background information, they were confronted by a hostile public. An offer to meet a week later with BHNA was turned down. The upcoming BHNA meeting had already been set aside "to build a case against RCA."[7] There was no doubt in the minds of many residents that the facility was not welcome (figure 8).

The opening legal battles in the dispute centered on the need for an environmental impact statement (EIS). Following its July 23 meeting, BHNA members wrote Rick Kimball urging that a statement be required because of possible adverse effects on land values, incompatibility with the character of the "predominantly residential and rural neighborhood" in question, and uncertainty over potential health hazards. A month later RCA argued that a statement was not necessary because the projected radiation levels were below every existing RF exposure standard, even the stringent USSR general population standard.[8]

To strengthen its case, RCA engaged the services of an RF bioeffects expert across Puget Sound, at the University of Washington, Bill Guy. Guy forcefully echoed RCA's sentiments. After stating that "there aren't any scientists or experts in this field in the entire world who would claim that biological effects could take place at the level of $3.28\mu W/cm^2$ (the estimated exposure level at the property boundaries)," Guy concluded, "Based on these data, a requirement for writing an environmental impact statement concerning health hazards or biological effects would be an unnecessary waste of time and

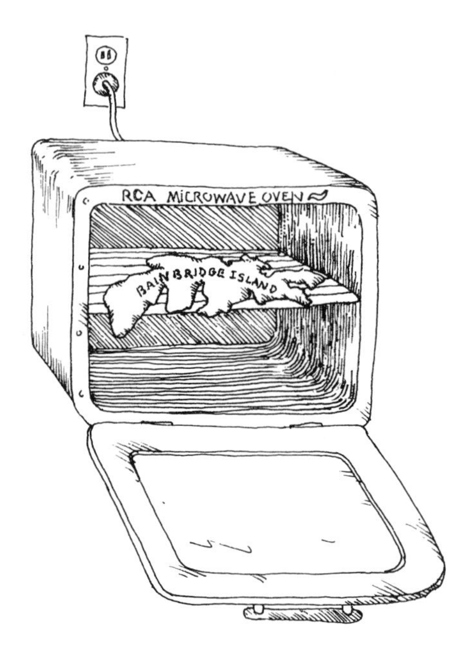

Figure 8
Cartoon by Tom Batey from the *Weekly*, a Seattle news magazine, September 9, 1981. Reprinted with permission.

funds." As a scientist Guy felt that "there is absolutely nothing in the scientific literature that would support claims of biological effects resulting from either short-term or long-term exposure to these extremely low levels." These were the facts as far as Guy was concerned. If they were properly understood and fairly presented to the public, there would be no reason to question the safety of the uplink facility.[9]

Guy's reassurances did not satisfy county officials. On August 21 RCA was notified that "an EIS will be required before processing of the application can proceed." Uncertainty over biological effects was cited as the main reason for taking this action. Furthermore RCA was put on notice that "of particular concern in the EIS will be an objective survey of current information on short- and long-term impacts on human health."[10] Bucklin Hill had won the first round in the fight with RCA, wrote the local newspaper.[11]

Hellmuth and his neighbors did not relax after their first apparent victory. They understood that opinion on RF bioeffects was divided and that the testimony given during the deliberations depended on the experts called. They had also been alerted to the possibility of conflicts of interest by reading the writings of Brodeur and his journalistic colleagues. Thus when they learned that Guy was to play a role in the formulation of the EIS they once again demanded action.

Washington's State Environmental Policy Act set guidelines for the preparation of the EIS. The final document had to be prepared "in a responsible manner and with appropriate methodology." All sides of important issues had to be weighed. Interested parties, such as RCA, could prepare the EIS, but the responsibility for content lay with the county: "No matter who participates in the preparation of an EIS, it is nevertheless the EIS of the responsible official of the lead agency."[12]

BHNA members doubted that an EIS drawn up under Guy's supervision could meet the requirements for objectivity and openness. Guy had already said on the record that there were no scientists and no scientific publications that could be cited to raise concern about the safety of the facility. Since his mind apparently was already made up, he did not seem to be an appropriate person to present a balanced and complete analysis of the RF bioeffects issue. BHNA members therefore once again asked Kimball to intervene. Their recommendation was that Guy either be dropped from the list of persons preparing the EIS or, if that could not be managed, his presence be balanced with experts who held other views.[13]

This time BHNA's recommendations were only partially accommodated. Guy was not made responsible for the bioeffects section of the EIS; his contribution was limited to the engineering portions of the document. The review of the bioeffects issue was contracted out to a local research establishment, the Battelle Human Affairs Research Center in Seattle. For the next six months Battelle's researchers, Guy, RCA, and other local consultants pieced together a document that would satisfy the requirements for a review while demonstrating the safety and environmental compatibility of the proposed facility. The product of their labors was made public on March 14, 1982, with local residents being given one month to comment.[14]

BHNA appealed to Kimball and asked for more time to respond. It had taken RCA more than six months to get opinions from its experts; BHNA wanted two additional months to contact other reviewers. State law prohibited the extension. By this time, however, Hellmuth had assembled a list of consultants who were sent copies of the EIS and asked to return comments to Rick Kimball. Many, including myself, did. By mid-April Hellmuth had assembled a volume of replies that nearly matched in size the original EIS.[15]

The process of getting the EIS prepared and reviewed sharply divided the RF bioeffects community. Battelle's section on human effects included a vigorous attack on the "journalistic coverage of the bioeffects of microwave radiation" in an attempt to discredit popular and some scientific literature that reported low-level effects. As one of the persons criticized in this section, Milton Zaret responded, accusing Guy of distorting the facts. Allan Frey also objected when he was grouped with a "radical minority" reportedly stirring up the trouble.[16] Guy, who had earlier been charged with conflict of interest in the University of Washington's student newspaper, the *Daily*, countered with his own view of the matter.[17]

In May 1982 controversy spread from scientific circles to the federal government when Hellmuth submitted a Freedom of Information request to EPA asking for the documents it had used to formulate its response to the EIS. EPA had gone on record on May 12, 1982, as supporting the accuracy and objectivity of the EIS: "The calculations regarding microwave radiation in the Draft EIS appear to be accurate and EPA agrees with the conclusion that, on the basis of the available health effects information in the literature, the exposure levels caused by the RCA earth station will not result in any adverse health impact on humans."[18]

The documentation released to support EPA's position consisted of two letters. The first, a short letter from the chief of the surveillance branch, Rick Tell, raised minor concerns about the calculated power levels. The second, a longer letter from the director of the Experimental Biology Division, Joe Elder, carefully documented a few of the many obvious biases contained in the EIS. Elder's final conclusion, which hardly supported EPA's public position, was that "the authors did a poor job of reviewing and citing the literature and reveal a lack of scientific insight into the complexities of the biological effects of microwave radiation."[19]

The controversy provoked by the shortcomings and biases in the EIS led to a change of plans. As early as November 1981 stories began to circulate that alternate sites were being explored by RCA for the uplink facility. Vascon Island, which lay south of Bainbridge Island, was one location mentioned; three undisclosed sites in North Kitsap were also reportedly under consideration.[20] On June 28, 1982 RCA's attorney wrote to Kimball, asking him to defer action on the Bainbridge Island site while a new application was submitted for use of one of the three locations in North Kitsap. Hellmuth and BHNA breathed a momentary sigh of relief, while their neighbors in North Kitsap prepared to continue the fight.[21]

Following the change in site the pace of the review process sped up considerably. Two weeks after the new plans were announced, the Kitsap County Planning Department reached a decision on RCA's application for an "unclassified public use permit/conditional use permit for a communication satellite earth station." Approval of the application was recommended, with only two minor stipulations, neither of which had bearing on the controversial health issue. The Planning Department accepted without question the assurances of safety given in the now slightly revised final EIS.[22] For the moment the RF bioeffects issue had been resolved in RCA's favor. The following week the proposal was to go before the county hearing examiner, Leonard Costello, and from there, if recommended, to the county commissioners for final acceptance.

As county hearing examiner Costello was charged with assembling evidence and making a recommendation to the county commissioners on a course of action. He had before him twenty-three documents submitted prior to a public hearing, held on July 25, 1982, the testimony of sixteen witnesses presented at the hearing, and twenty-nine additional pieces of evidence submitted after the hearing. The written evidence ranged in size from single-page, hand-written letters to the draft and final EISs.

Costello's reasoning in reaching a decision closely paralleled the arguments RCA had used to assure safety. He acknowledged that some members of the public opposed the facility and he admitted that there was disagreement within the scientific community on the health hazards issue, in part because critical research data were lacking. But Costello felt that there also was evidence to suggest that the facility did not violate the county's zoning ordinances or comprehensive development plan. The facility did not, in Costello's view, compromise the rural designation for the land being used or pose "a substantial risk to human health."[23]

It is difficult to determine how Costello reached his conclusion on the health issue. Of most importance to him seems to have been the fact that the radiation levels being contemplated were "significantly lower than the levels allowable under the most restrictive of standards in effect." Costello apparently accepted these standards as his definition of safety and concluded that lower exposed levels were safe. Unlike the public he made no effort to assess the validity of these standards or to question their applicability to the Bainbridge Island case. He also made no judgments about the validity of the EIS. His concluding remark on safety simply noted, "Given the evidence presented, it does not appear to the Examiner that this proposal presents a substantial risk to human health due to the radiation generated during operation."[24]

Costello's recommendation to approve RCA's application left one final level for appeal, the three commissioners for Kitsap County. They could be guided by Costello's advice or reach their own conclusion. They chose the latter action. On October 18, 1982, a three-page memorandum drafted by Commissioner John Horsley was adopted unanimously by the three. The memorandum refused RCA's request, citing both land use and health considerations.[25]

The land use issue came to focus on one basic question: whether the use proposed by RCA was compatible with the rural-forestry classification specified in the county's long-term land use plans. Costello felt that RCA could make its facility compatible with surrounding land; the commissioners accepted the feeling expressed by community members that the proposed use was not compatible. The public feared that granting permission for one industry to build would set precedents that ran counter to the intended use of this rural region. The commissioners agreed.

The rejection for health reasons was based on two considerations. First, the commissioners were concerned about the "adequacy of the EIS." They did not accept the document at face value and in fact seem to have been influenced by its critics. Second, and probably the point that will prove more controversial as the commissioners' decision is appealed, they based their assessment of the safety issue as much, if not more, on public perceptions as on scientific data: "The important thing here is not the scientific debate as to the extent and degree of present or long-term risk. What is important from a public policy perspective is the degree to which a broad segment of the affected populace perceives that they and their children are at risk."[26]

A literal reading of the last judgment leaves open the possibility that public fears themselves could be used to block any building project, regardless of whether the fears are justifiable; however, this seems not to be the intent of the argument. Rather the commissioners seem to be saying that in situations where the safety issue cannot be resolved fully, as indeed it cannot be in the RF bioeffects area, public opinion must be taken seriously in judging the significance of risk because the local residents, and not industry, are the ones who will suffer any adverse consequences. In such situations it becomes irrelevant that "RCA may be willing to take [a risk] based on what they know and based on what they have to gain."[27] They are not the party that stands to lose. It is the public that is potentially at risk, so it was their willingness or unwillingness to be subjected to risk that weighed most heavily on the commissioners' minds.

The commissioners' ruling has not ended the debate. RCA contested the ruling and their appeal is still pending as of early 1984. In addition the commissioners' ruling did not clarify the bioeffects issue. They did not rule that the facility was unsafe; they simply argued that the degree of safety was not well enough established to warrant granting permission to build. Their ruling did, however, add one important and neglected ingredient to the microwave debate: it drew attention to the question of significance.

Commissioner Horsley set out in his memo criteria for judging the acceptability or unacceptability of potential risk. In doing so he established grounds for judging the significance of a potential health threat. No prior efforts at decision making had been as specific on this issue. Costello had not defined a "substantial risk to human health." The EIS gave no definition for "harmful to human health." ANSI's standards have never set criteria for judging "harmful to human health," even though explicit and implicit assumptions about safety are made. Congress did not provide the executive with guidelines for judging "protection." Horsley was the first person in the RF bioeffects field to venture publicly into the area of assessing the significance of risk.[28]

Horsley's assessment could be significant for the microwave debate if it results in some sustained discussion of risk. A clearer definition of risk would be useful in standard setting. It would help the public understand better just how safe or hazardous a proposed facility is judged to be and whose criteria were being used to evaluate risk. The addition of greater clarity might in turn eliminate some of the anxieties that have led not only to the confrontation on Bainbridge Island but to a rash of personal injury suits.

Alleged Personal Injury

Personal injury claims center on health problems alleged to have been suffered. There are basically two legal avenues for pressing claims of personal injury: workers' compensation and personal damage suits. Both have been used by persons who have claimed injury as a result of exposure to RF radiation.

The legal basis of workers' compensation claims are federal, state, and local laws that establish rules for compensating workers for injuries suffered on the job. For example, on January 5, 1978, Joseph Kerch, a fifty-eight-year-old pilot who experienced hearing and vision loss after years of flying, filed a claim with the U.S. Department of Labor seeking compensation for injuries. Air America, Inc., had been his employer. Kerch claimed that his vision loss was the result of cataracts caused by exposure to radar while working as a pilot. He was going blind, allegedly as a direct result of his work experience. If this were true, he was due compensation under federal law.[29]

The resolution of Kerch's claim depended on the answer to one central question: Were the injuries, principally the cataracts, caused by work-related factors? If they were, then Kerch qualified for compensation; if not, he did not qualify. The task faced by Kerch's lawyer, Matthew Shafner, was making connections between work conditions and the injuries. The defendant, Air America, Inc., had to show that such connections could not be made. Crucial to both cases was the validity of Milton Zaret's controversial microwave cataract theory.

The evidence supporting Kerch's claim followed a familiar pattern. During his early career his vision was normal. While working for Air America, his near vision weakened. By the mid-1970s he began to notice a halo effect around objects. In 1977 his ophthalmologist diagnosed his problems as cataracts and referred him to Zaret. Zaret characterized the injuries as typical radiant energy cataracts. Since Kerch's main exposure to RF radiation had been while flying, his cataracts were presumably job related.

The evidence against the claim, presented in part by ophthalmologist Dorothy Leib, contested Zaret's diagnosis. Leib agreed that microwaves could produce unique cataracts but did not believe that Kerch's cataracts were microwave cataracts. Zaret described the cataracts as posterior capsular; Leib did not. If her diagnosis were correct, Kerch's cataracts could not be traced to a unique work place condition (exposure to RF radiation), thus weakening the claim. Kerch also had other medical problems that had contributed to the general decline of his health. He had been diagnosed as once having a detached retina and was being treated for gastrointestinal cancer and a rare tropical disease. Thus there were many complicating factors in any diagnosis.

Kerch's claim was settled out of court in early 1982 for $30,000, leaving the problem of cause unresolved.[30] The settlement did not establish connections between the claims and payment. Kerch was not being compensated for RF injuries, nor was he being denied compensation. No legal precedents were set. However, legal precedents had been set already in another case that was being appealed when Kerch's claim was settled. The injured party in the other case, Samuel Yannon, had died. The claim for workers' compensation had been filed by his wife, Nettie Yannon, on January 11, 1975.[31]

Yannon's occupational exposure to RF radiation began in 1954 when he was hired as a radioman by New York Telephone. Among other duties he was responsible for tuning about two dozen low-power RF transmitters located in the Empire State Building. Two or three times each working day he entered the room housing the transmitters and spent about twenty minutes adjusting them, working sometime from behind the units, sometimes in front of them. In 1968, after fourteen years on the job, he began to experience health problems and was placed on disability leave. He was fifty-seven years old. Three years later he was retired from New York Telephone for health reasons. He died three years later.

No one doubted that Yannon died from health problems that first appeared when he worked for New York Telephone. He was in good health until about 1968. After 1968 his health declined rapidly: his hearing and vision failed, he lost his coordination and memory, and his weight dropped from the 180s to under 100. The main issue that needed to be resolved in settling the claim was whether the health problems were actually work related. Nettie Yannon, Sam Yannon's physician Alfredo Santillo, and Milton Zaret argued that there was a connection between the work environment and the health problems. New York Telephone representatives, Sol Michaelson, and a colleague of Michaelson's at Rochester, Robert Hendon, attributed the death to non-work-related causes. The decision on the claim depended on which of the experts was believed.[32]

From a scientific point of view the claim had many weaknesses: The exposure levels were not known, the equipment Yannon had adjusted was no longer in operation so direct measurements could not be made, and it was impossible to recreate his exact work pattern. Did he follow suggested safety rules, turning the antennas off when he worked in front of them? The principal expert supporting the claim, Milton Zaret, was himself under attack by the time hearings took place. Zaret's scientific expertise was limited to eye problems; this case involved much more. The scientific community had never accepted the existence of effects under such ambiguous and controversial conditions. They were not likely to testify in support of this claim.

Law and science do not operate in the same way, however. Only two conditions had to be met to establish legal proof: the disease Yannon contracted—"microwave radiation sickness"—had to be a recognized occupational disease, and he had to contract that disease as a direct result of his normal work experience. Workers' compensation covers health problems experienced as a direct consequence of normal work conditions. If these two conditions could be satisfied—contracting a recognized occupational disease as a direct consequence of normal work conditions—the compensation would be awarded.

The criteria for legal proof worked in the Yannons' favor. The legal experts who were the final arbiters in the workers' compensation hearings accepted Zaret's contention that microwave radiation sickness is an occupational disease. (Zaret's scientific colleagues had not accepted the evidence he had published over the years as proof that humans were being injured by RF exposure.) Once it was accepted that others had suffered from microwave radiation sickness under exposure conditions similar to Yannon's, making the causal link required easily followed. If Zaret had uncovered examples of the disease, he must also have shown the causal links.

Zaret's pivotal role in this case is more than evident in the final rulings. The three-member Workers' Compensation Board panel who reviewed the original award, made in June 1980, ruled on February 28, 1981, "Upon review the Board Panel finds based on the entire record and particularly the testimony of D. Rieu [a radioman], and Drs. Santillo and Zaret, that there was a direct causal relationship between decedent's [Yannon's] exposure to microwave radiation during his employment and his subsequent disability all of which resulted in his death."[33] Following another appeal, on May 6, 1982, the New York State Supreme Court Appellate Division ruled that "claimant's leading expert, Dr. Milton Zaret, provided the Board with ample evidence of the existence of a disease identified as 'microwave or radiowave sickness.' Dr. Zaret's own studies, including those performed for the United States Government, and excerpts of reports from the Warsaw Conference of 1973 which documented the diagnosis of such a disease in other countries, substantiate this conclusion. The Board was entitled to credit his testimony and that of other experts supporting this view."[34] The scientific community may have had doubts about Zaret's work; the legal community did not.

Assuming they are not overturned, these rulings lead to an awkward situation. The legal community has gone on record in the Yannon case as recognizing microwave radiation sickness as a disease. With few exceptions, principally Milton Zaret, the scientific community has not. Moreover Zaret has himself paid most attention to possible eye problems and not generated experimental evidence to verify his more general claims. As a result the evidence supporting the legal position is thin at best and perhaps nonexistent.

The lack of scientific evidence to support the legal decision in the Yannon case should not be interpreted as indicating that the decision was wrong. Had the burden of proof been on the other side, the evidence would have been equally as weak. The evidence linking Yannon's illness to causes other than RF exposure was as weak as the evidence making ties to RF exposure. Therefore it is impossible to determine which of the two positions, the legal or the scientific, is correct. And this situation will not change until more attention is given to such cases by the scientific community.

The willingness of lawyers to help clients test the RF bioeffects issue is evident not only in workers' compensation claims but in personal injury suits. Relying on her legal counselors, Nettie Yannon pursued this second course of action in 1976 when her workers' compensation claim had temporarily failed to secure the hoped-for decision. She sued RCA, the company that manufactured the equipment her husband had serviced, for breach of warranty and negligence, claiming that RCA had not acted responsibly to protect her husband from known dangers. Their irresponsibility, she claimed, led to his death, and she asked for $3.5 million in damages.

Two months after the New York State Supreme Court Appellate Division upheld the workers' compensation ruling in the Yannon case, a state judge dismissed Yannon's personal injury suit on technical grounds. Under state law the warranty and negligence suits had to be filed within three years of the injury's being sustained. Since Sam Yannon's signs of ill health appeared in 1968, the statute of limitations ran out in 1971. Nettie Yannon claimed that her husband was mentally impaired by 1968 and could not file the suit. The judge who heard the case disagreed, ruling that Sam Yannon was sane in 1968 and could have brought the suit. Since he did not, a motion to dismiss was granted. This time it was Nettie Yannon who appealed and is still waiting.[35]

Had the Yannon suit not been dismissed in 1982, six years after it was filed, it probably would have followed the same legal paths as did a suit filed on March 17, 1977, by a navy radar repairman, Robert Engell. Engell and a friend, Donald Cadieux, worked at Quonset Point Naval Air Station in Connecticut. Both reportedly spent many hours each day near transmitting RF equipment. The radar they repaired sat on waist-high repair benches, placing it at times only a few inches away from their abdomens. Cadieux died in October 1974 at age thirty-one, allegedly of complications stemming from pancreatic cancer. He had spent eight years at Quonset Point. Engell worked at the same facility for twelve years, prior to his retirement in 1973, and was suffering from what was reported to be pancreatic cancer a few years later.[36]

The appearance of two rare forms of cancer in two relatively young men who had the same work experience raised the strong suspicion of a work-related health problem. Engell tried to argue this point through a workers' compensation claim but was not successful. Thus he decided to press a personal injury suit with the help of the same lawyer who later took on the Kerch case, Matthew Shafner, and New York attorney Marc Moller. The defendants named in the suit were the electronics corporations that manufactured the equipment Engell had repaired: International Telephone and Telegraph, Raytheon, Varian Associates, Teledyne Ryan Aeronautical, Rockwell International, Bendix, and General Dynamics. The plaintiff, Robert Engell, asked for $4.5 million in damages and a jury trial.[37]

The dollar amount of the Engell suit was well in excess of the award requested in workers' compensation claims, and exceeded Nettie Yannon's $3.5 million suit. But as important as the money were the precedents. If Engell won his suit, the way could be opened for tens or perhaps hundreds of similar actions. Newspaper accounts suggested "that between 12 and 15 of Engell's coworkers . . . have also developed 'abdominal problems' or cancer."[38] Donald Cadieux's wife, Louise, followed Engell's lead in June 1977 and sued the same electronics corporations for $4.5 million.[39] Joe Towne was actively circulating information on legal actions taken by members of the Radar Victim's Network and urging others to take similar actions.[40] From the corporate point of view, the Engell case represented the thin edge of the wedge and had to be fought. Engell's lawyers were equally as determined to get what they considered to be just compensation for their client.

The Engell case never went to trial. Between March 1977 and December 1982 the two sides sparred with one another in the fact-gathering stage. Engell's lawyers, fearing their client would not be alive at the time of the trial, petitioned to have his deposition recorded on video. Their petitions were granted, and five videocassettes with Engell's testimony were finally put on record on September 14, 1977. The plaintiffs drew up a list of eighty-four questions submitted to the corporations for answers. The defendants repeatedly asked for additional time to answer and drew up their own lists of questions for Engell. Depositions (preliminary testimony by potential witnesses) were entered by both sides, with some of the potential witnesses being subjected to extensive cross-examination. Zory Glaser, an RF bioeffects expert who worked for NIOSH and later BRH, spent five days answering questions put to him by the lawyers for industry and an adviser, John Osepchuk. Osepchuk also served as a preliminary witness for industry. He was to be joined later by Justesen and Michaelson.[41]

Toward the end of 1982 the fact-gathering stage drew to an end and the trial date at last seemed near. Then in November 1982, four of the companies left in the case (the suit against Bendix had been dismissed on January 15, 1980) reached out-of-court settlements with Engell. The company remaining, ITT, stayed with the case through December and the selection of a jury. A new trial date was set for January 5, 1983. Before the trial date arrived, ITT also reached an out-of-court settlement.[42] The terms of the settlement were not made public.

The settlement provided a convenient means for resolving a difficult legal matter. Neither party lost; neither won completely. Engell received an award that was presumably large enough to satisfy him that he had been compensated for his injuries. In exchange he had to abandon his attempts to prove in a legal sense that his employers had been responsible for his ill health. The corporations had to pay a settlement but were not found guilty of negligence. No one could cite this case as a precedent in the future. Other suits brought for similar reasons would have to go through the slow, expensive pretrial process followed in the Engell case. To this extent at least industry had held at bay the wave of suits that could follow the return of a verdict supporting the plaintiff. As a legal tool the out-of-court settlement had advantages.

But as a tool for helping to settle the dispute that had led to the suit, the out-of-court settlement was of little use. Five and a half years of legal maneuvering produced no conclusions about the cause of Engell's illness or the extent of industry's liability. The same was true in the Kerch case and in the majority of other legal actions involving RF bioeffects, most of which have been resolved out of court.[43] Because experts in government and science have not been able to resolve disagreements about the alleged danger of RF exposure, legal actions have been looked to as a way of getting answers. But answers have not followed. The law has provided settlements for individual cases, but with the sole exception of the Yannon case it has not reached any conclusions about alleged human injury.

Closing the Circle of Indecision

The inability of legal proceedings to cool down the microwave debate is in some ways difficult to understand. To some extent lawyers are better equipped to handle ambiguity than are scientists or governmental officials. Lawyers are trained to dissect arguments and, through cross-examination, to sort out fact from fiction. On more than one occasion, scientists, whose testimony has been important in particular cases, have been subjected to intense questioning about their methods, data, results, and values. The depth of inquiry in the courtroom usually far exceeds the rigor of scholarly dialogue or the give-and-take of the government hearing.

But as rigorous as the questioning of lawyers can be, it cannot provide clarity where there is none. Some have tried. In a suit brought in 1975 against Litton Industries by Helen Mulhauser and Agnes Ryan for alleged injuries stemming from a microwave oven, Zaret was called on by the plaintiffs to supply evidence on possible microwave cataracts. Defense attorney Richard Bennett spent one full day questioning Zaret's diagnosis and his credentials as a scientist. Articles published by Zaret were discussed, line by line. His navy work and Warsaw paper on the alleged Finland problem were worked into the questioning. Every criticism of Zaret's work that could be located was used, including confidential reviewer correspondence from the *Journal of the American Medical Association* rejecting an article Zaret had submitted for publication.

In his questioning Bennett attempted to do something that the scientific community had neglected; he tried to understand and put on the record Zaret's complete theory of microwave cataracts. Zaret had to define terms, describe his observations, explain diagnoses, clarify sources, and provide other details on his work. The resulting deposition provides what is perhaps the clearest description in print of the controversial microwave cataract thesis.[44]

Zaret's responses raised many questions about his work. He could not recall sources, was vague about evidence, and sometimes appeared to contradict himself. He admitted that his work was virtually unreplicated and in most ways unique. Even the terms he used to describe microwave cataracts were his own. And his clinical procedures definitely opened the door to charges of circular reasoning. His diagnosis of a microwave cataract depended on some knowledge of prior history, while histories in turn could raise suspicion of microwave cataracts.[45]

Throughout the questioning, however, Zaret steadfastly maintained that he could see changes in the posterior capsule surrounding the lens that stemmed from RF exposure.

> Bennett: It has a very distinct pathonomic [sic] evolution, a microwave cataract?
>
> Zaret: Yes.
>
> Bennett: It has roughening of the lens capsule?
>
> Zaret: In the early stages. Not every ophthalmologist would agree with that. Not every ophthalmologist can see it.
>
> Bennett: How do you determine the presence or absence of roughening?
>
> Zaret: When you play the light beam along a very fine beam [line?] and look with the proper magnification and the right angle, you see it.
>
> Bennett: What is it? . . .
>
> Zaret: It looks like the surface is puckered and rough. . . . [It] also looks like it is thick.[46]

This simple I-can-see-it-even-if-you-cannot argument was the most difficult to reject. Other ophthalmologists could always be brought in to testify that they could not see the roughening and thickening that Zaret saw. As long as this was the case, there was always a chance that Zaret, and not his critics, would prevail in court, as he did in the Yannon case.

Whether because of uncertainty about the outcome or other factors, *Mulhauser-Ryan v. Litton Industries* was settled out of court. This time a new twist was added to the accommodation reached. On January 9, 1976, Mulhauser and Ryan petitioned the court to have the case dismissed, "with prejudice, and without costs to any of the parties." The major reason given was that following the deposition, Zaret had refused to supply further information and had withdrawn his support for the case. The plaintiffs concluded, "We now believe that Dr. Zaret had no scientific basis for his claimed finding that we had microwave eye injuries and cataracts caused by exposure to leakage from a microwave oven. We believe Dr. Zaret has misled us and consequently caused us to expend enormous time and effort in futile and unnecessary litigation."[47] Bennett had accomplished his major objectives; he had won the case for Litton Industries and legally (not scientifically) discredited Zaret's testimony. Whether any cash settlement went along with the agreement to dismiss with prejudice has never been made public.

The difficulties Zaret encountered when called on to defend his ideas in court have been experienced by witnesses on the other side as well. Herman Schwan's lengthy testimony at the Rockaway Township Home Box Office hearings is ample demonstration. Schwan was called to provide expert testimony on RF bioeffects. As a leading researcher in the field, he seemed eminently qualified to bring rationality and solid evidence to bear on the impending zoning decision. Under rigorous and sometimes hostile questioning, he fared no better than Zaret.

The major problem for experts in legal settings is a dichotomy that exists between the objectivity of their science and the subjectivity of their testimony. Schwan was able to provide the Rockaway Township Board of Adjustment members with factual information on his own work. His objectivity as a scientist was assumed, just as it must be assumed until proved otherwise that Zaret is reporting objective clinical observations when he testifies in microwave cataract cases. When called on to provide information on other work in the RF bioeffects field and on the state of general knowledge about RF bioeffects, however, Schwan was being asked to go beyond facts to opinions. Admittedly the opinions he supplied were those of an expert in the field, but they were still opinions.

The Rockaway Township Board members understood that they were under no obligation to believe everything Schwan or any other expert said. But it still remained for the board, or in other cases, the judge or jury, "to decide how much weight it wishes to give to [the] testimony."[48] The weighing process is vital to the legal process. The microwave debate has arisen because the experts who testify disagree; someone has to decide who to believe. This is the burden that has been placed on judges, jurors, and public officials when they are asked to confront the RF bioeffects problem in hearings and in court. When the divisiveness of chosen experts is added to the impossible task of finding solutions that have eluded the experts, it is not difficult to understand why the out-of-court settlement has been so popular. It is the legal community's way of avoiding decision making. Congress can pass the RF bioeffects problem along to the executive. The executive can drag its feet. Scientists can always ask for more time and money to gather more information. The media can popularize without facing up to the facts. And the legal process can put out small fires without getting burned by the main inferno. The circle of indecision is complete, the underlying problem remains.

10

Science and Values

In [the last] two decades, the United States has received enormous amounts of scientific and technical information and advice from the scientists and engineers of the country. The information is almost always technically correct and thorough; it is almost always given with the intention of solving or mitigating the problems sketched [out]. The paradox is simply this: How have we gotten into so much trouble while getting so much well-intentioned and correct technological advice?—Martin Perl, *Science* (September 1971)

Many specific reasons have been advanced to explain why the debate has been ongoing for several decades. In general all can be reduced to either the failure or the inability of the institutions and persons engaged in the debate to confront the basic problems that arise when technology, science, and values are in conflict. Congress and the executive have not been able to make decisions about regulating RF technology. Scientists have not critically analyzed their role in policymaking. The mass media have not carried out their obligation to educate the public in a full and responsible way. The legal system has been unable to provide clarity where none exists.

It is tempting to conclude this study with a call for increased efforts to make the system work better. My own feeling, after observing the debate for a number of years, is that the current system could be made to work better but only if some fundamental changes are made in basic assumptions about appropriate problem-resolution procedures. Of most importance in this regard is a need to reconsider the relative importance assigned to two critical components of all problem solving: science and values. I believe that the procedures used for dealing with the microwave debate and similar ones have placed too much emphasis on science and too little on values.

Science and Its Limits

There is a seductive plausibility to the argument that a detailed scientific understanding of RF bioeffects is essential for decision making. After all, how can RF standards be set, to focus on the most controversial policy issue, if bioeffects are not fully understood? The purpose of standards is to protect human health. In order to protect health, we need to know how RF radiation affects health. Unquestionably such knowledge is best gained through science. Objective decisions require objective information; objective information is gained through science.

As logical as this line of reasoning seems, it does not necessarily lead to the conclusion that problem solving is best begun by assembling more scientific data on RF bioeffects, as has so often been argued. The RF bioeffects problem is complex and needs to be clarified before experiments are undertaken, especially if the results are going to be used for establishing policy. Experiments have to be planned with an eye toward generating the specific information needed for making decisions. Moreover it is not true that detailed scientific information, especially about RF bioeffects, is as critical to problem solving as is usually assumed. The $10mW/cm^2$ figure that Herman Schwan proposed in 1953 as an estimate of presumed safety can be used for making many important decisions. From the mid-1950s on few persons denied that exposure above $10mW/cm^2$ started to get into the gray zone of possible hazards, with $100mW/cm^2$ being accepted as the level at which health effects are expected. Was anyone being routinely exposed to such levels of radiation? From a public policy standpoint, this would logically seem to be the first question to ask. If they were not, then the inquiry process could turn to the question of the adequacy of the initial estimates. If they were, more immediate steps would be called for to protect persons from hazardous exposure situations.

As logical as this first step might seem, it was not incorporated into problem-solving efforts in any major way until the mid-1970s when EPA began its environmental monitoring program. Prior to that time the military ran spot checks on a few facilities but never within the context of an overall assessment of the total exposure received by its personnel. Random environmental monitoring in industry began in the early 1970s with the establishment of OSHA and NIOSH, but again not within the context of general assessment. As a result even today it is not possible to give a complete answer to the basic question of whether people are being exposed routinely to more than 10mW/cm^2 of RF radiation, and if so, how many and under what circumstances. When Congress tried to assess the possible extent of exposure from RF sealers and VDTs in 1981, it could not get definite answers.[1]

The folly of making decisions without information on exposure is well evidenced by episodes such as Project Big Boy, the secret investigation undertaken during the Pandora years to discover whether exposure to shipboard radar produced adverse health effects. The project failed because it was discovered after the fact that above- and below-deck exposure levels were not as different as expected. Had the navy undertaken a routine monitoring program when it first became concerned with RF exposure in the early 1950s, actual exposures on shipboard would have been known and the experiments could have been designed accordingly. Much the same conclusion follows for every other personnel survey that has ever been conducted.

But even the lack of detailed exposure data is not a major obstacle to further planning. For example, it has always seemed plausible to assume that those who work in the vicinity of RF equipment are exposed to considerably more RF radiation than is the general population. How much more is still not known since work place averages have not been determined. We know from EPA's environmental monitoring program that general population exposure on an average falls below 1μW/cm^2 (10,000 times below 10mW/cm^2), with maximum exposure occasionally reaching as high as 100μW/cm^2. If work place exposure is significantly greater, then it could fall in the low mW/cm^2 range or above, a figure borne out by the isolated monitoring that has been conducted in work place settings.[2] These rough estimates, which have commonly been accepted from the mid-1950s on, provide more than enough information for planning subsequent actions.

The fact that general population exposure to RF tradition is at very low levels for long periods of time strongly suggests that the focus of scientific experimentation should be chronic, low-level studies. High-level (10mW/cm^2 and above) thermal experiments may be of interest from the standpoint of pure science, but they have little relevance to the general population. This conclusion is not controversial. On the contrary, it is a necessary one that follows from the rationale used to set the $10mW/cm^2$ standard in the first place. This standard was adopted because there was supposedly compelling evidence to prove that thermal effects did not occur at lower levels. If this evidence were truly compelling, it follows that thermal experiments cannot provide information on effects that might occur in general populations at lower exposure levels. Thus, the only reasonable way to test the premise that the general population is safe is to undertake extensive chronic, low-level exposure experimentation.

As logical as this conclusion might seem, it has never had the impact it should on RF bioeffects research. Why? Principally, I argue, because insufficient attention has been devoted to research planning. Such planning has been conducted in a haphazard fashion, often behind closed doors, relying on only a limited number of experts, and with very little attention being paid to the planning process itself. Great care is usually taken to ensure that individual experiments are undertaken appropriately by turning to RF bioeffects experts for reviews. However, planners have never sought advice from persons who study science policy and who could, on the basis of extensive experience with other fields, comment on whether the overall research program being pursued is appropriate. The result of this situation is that after thirty years of more or less constant research effort, there are still gaps in the scientific record.

Before research can be undertaken, it must be planned; research planning requires more than input from scientific experts in the proposed field of research. In-field experts have two shortcomings: they are committed to particular lines of research and usually support these lines when they give advice, and they have seldom been trained to study science in its broadest contexts. Policy experts from outside science (sociologists, political scientists, economists, historians, philosophers, and others) are not committed to specific fields of scientific research and can make rigorous judgments about what can and should be done. Over the last twenty years the U.S. government has spent billions of dollars building the social sciences in the belief that they are useful. Certainly in areas such as the microwave debate, they should be put to use in advance of allocating research funds for particular research projects.

The need to draw a broader range of experts into the microwave debate is also apparent from a second conclusion that follows from the general exposure data that have been available for many years. If it is true that most people are not routinely exposed to more than a few $\mu W/cm^2$ of RF energy, then it would seem to follow that RF standards could be set well below $10mW/cm^2$ without causing much economic hardship. If EPA's measurements are accurate, it would appear that very few RF facilities expose surrounding populations to more than $100\mu W/cm^2$ levels. This means that the economic cost of a $10\mu W/cm^2$ standard would be limited. Certainly this was even more the case twenty years ago when standard setting began in earnest.

The conclusion that general population standards could have and probably still could be set well below $10mW/cm^2$ or even $1mW/cm^2$ without causing serious economic burdens is relevant to decision making because it could alter the course of decision making. On the basis of this conclusion policymakers could have decided to set standards at the limit of economic hardship rather than at the limit of biological effects. Efforts to define the limits of economic hardship would have required not only engineering studies but social and economic analyses, thereby giving a totally different and more socially responsive tone to the problem-solving effort. More social responsiveness might have lessened public anxiety, thus eliminating a major cause of controversy. Had decision making been set in broader contexts from the start and followed a logical problem-solving path, it is possible that the microwave debate could have been avoided, at least in the public sphere.

The principal reason why more comprehensive approaches to problem solving have not been adopted can be traced to the uncritical adoption of narrow scientific methods. Our faith in science and its ability to provide objective information raises the hope that it can solve all problems; hence we give way to the temptation to collect more scientific information without stopping to wonder whether the problems being addressed are amenable to narrow scientific solutions. Science is an efficient tool for discovering facts about nature. Making decisions about how best to exploit and regulate RF technology requires more than factual information about nature. Such decisions require information about economics, politics, law, history, social contingencies, and much more, information beyond the expertise embodied in science.

Values

The overemphasized hope that problem solving can be objectified by turning to science has had one additional impact; it has led to a failure to recognize the importance of values in decision making. The reason for the failure is obvious. If decision making can be objectified by turning to science, then values can be ignored because presumably they can be eliminated. The fallacy inherent in this argument is equally as obvious. Try as they might, policymakers have not been able to eliminate values from the decisions they have made. To illustrate this conclusion, one need look no further than the latest revision of ANSI's RF standard, which was adopted in 1982 (ANSI C95.1-1982).[3]

The scientific bias of all of ANSI's RF standards activities is clear. From the early 1960s on ANSI's standards committees have looked at and discussed only one type of evidence, scientific evidence. The latest revision of C95.1 is no exception. The only literature cited is scientific literature on RF bioeffects. The only issues discussed in any detail are scientific issues. There is no evidence from existing documentation to suggest that either C95.IV or C95, the committees that respectively drew up and adopted C95.1-1982, has ever had a serious or sustained discussion of any topic other than science.

ANSI's total preoccupation with science could be accepted if values did not enter into their deliberations, but values have had a bearing on the standards they have set, and these values have had a profound impact on the microwave debate. Some of the values are obvious, such as the preoccupation with science itself. Others are less obvious but no less important, as can be illustrated by taking a closer look at some of the decisions made during the formulation of C95.1-1982.[4]

ANSI members have always derived RF standards by working from a range of known danger to a point of presumed safety. They knew as early as the 1950s that exposure at 100mW/cm^2 could cause overheating and damage. The challenge posed to scientists for the purpose of setting standards was to discover how far below that level standards had to be set to avoid causing injury. They attempted to find the lowest level at which effects could be found and then set standards a presumed safe distance (usually a factor of ten) below that level.

This approach to setting standards is not the most conservative (defining conservative as offering maximum health protection) one that could have been adopted. C95 members could have spent their time discussing how low standards could be set without infringing on economic development instead of figuring out how high standards could be set to avoid injury, an approach that probably would have resulted in considerably lower standards. Standards could not be set much higher however without leading to the possibility of hazards, a fact C95 members recognized, even if they did not understand its anticonservative consequences. Persons relying on C95.1-1982 are warned that "exposures slightly in excess of the radio frequency protection guides are not necessarily harmful, however, they are not desirable and should be prevented whenever possible."[5]

The lack of conservatism assumed in C95.1-1982 is aggravated by its strict scientific bias (a value). Within the upper-limit context, greater protection (conservatism) is achieved by lowering standards in response to new information on effects that may occur at lower levels. Any factor that serves as an obstacle to getting new information into the standard-setting process could potentially militate against lower standards. One major obstacle that has consistently been used to exclude information from U.S. standard-setting efforts is the selected use of scientific rigor.

In the past U.S. policymakers have used supposed scientific rigor to exclude most East European research on RF bioeffects from their deliberations. The framers of C95.1-1982 are no exception. "Some effects reported in the Eastern European literature were discounted because of questionable control procedures and lack of information on environmental parameters and physical measurements." An effect known as calcium efflux, which has been shown to be modulation specific at low exposure levels, was ignored because it could not be tied to any adverse health effects.[6] The data used to set C95.1-1982 had to overcome the obstacles of "demonstrability . . . , relevance, reproducibility, and dosimetric quantifiability."[7] Data that failed to meet these criteria were not used.

What would have been the consequences of lowering the requirements for judging scientific rigor? Work of doubtful scientific quality would have made its way into standard setting and reduced the reliability of the standard. Insofar as the questionable data supporting the standard could be criticized, so too could be the standard itself. There can be no doubt, however, that if some of the questionable data were used, C95.1-1982 would have been set at a lower level; in other words sacrificing scientific rigor would have resulted in a more conservative policy toward health protection. In choosing to stay rigidly with scientific rigor (a value judgment), C95.IV members again shied away from pursuing a more conservative policy.

That such decisions have been based on underlying values is more than apparent from the selective way in which scientific rigor has been used in C95.1-1982 to quantify exposure. Previously C95.IV members were content to express standards in terms of the amount of energy arriving at the skin surface. The basic mW/cm^2 measure is simply an expression of the amount of energy per unit of area. Scientists have known for many years that some of the energy that arrives at the skin surface is not absorbed into the body. Thus surface exposure is not an accurate measure of the amount of internal energy that produces bioeffects. To compensate for this inaccuracy, the new standard has been written in terms of the actual amount of energy absorbed, or "specific absorption rate" (SAR).[8] One consequence of the added precision has been relaxation of the new standard at some frequencies, based on the fact that at these frequencies RF energy is not readily absorbed (figure 9).

The studies that led to the formation of the SAR concept have also demonstrated that the body can concentrate RF energy, producing localized hot spots with energy levels that equal or exceed surface levels (figure 10).[9] Thus the inaccuracy of using surface energy levels to regulate exposure cuts two ways. Internal energy levels can be either higher or lower than surface exposure, depending on where within the body measurements are made. Whereas the knowledge that internal energy levels can be lower was used to raise standards, knowledge that internal energy levels can be higher was not used to lower standards. The precision gained by use of the SAR concept has been used selectively (a value judgment) in setting C95.1-1982.

The reasoning used by C95.IV members to dismiss hot spots from their deliberations is difficult to accept. They contend that whole-body averages can be used for calculating standards because such averages take hot spots into consideration. Average whole-body exposure is calculated by weighing hots spots and cold spots together to determine mean and average exposure. That for every above-average exposure spot within the body

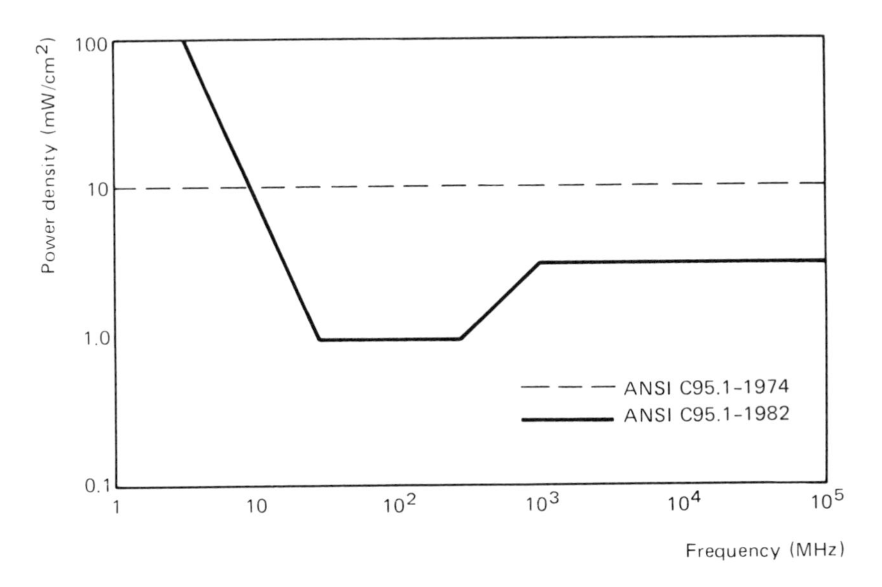

Figure 9
Comparison of the exposures allowed under the old and new ANSI standards.

there is a corresponding below-average exposure spot is of no consequence at all if the above-average exposure produces damage. The average whole-body momentum delivered by a 1 ounce bullet traveling at 500 feet per second is about one hundred times less than that delivered by a 200 pound football player running at 12 miles per hour. This fact would offer little consolation if the point of impact of the bullet were the heart. That whole-body SAR may be within acceptable limits is also of little consequence if the above-average exposure occurs in a place of vital importance to well-being.[10]

Why was the SAR concept used selectively in setting C95.1-1982? The answer to this question is obvious. The use of SAR data regarding hot spots in setting standards would have been a conservative move. The use of average whole-body SAR data, of scientific rigor, of the upper-limits principle, was not. C95.1-1982 is, if nothing else, consistent. When given an alternative, its authors shied away from greater conservatism, not toward it (a value judgment). As simple a step as increasing the margin of safety was rejected because "none of the members of the subcommittee offered an argument" in favor of this step.[11] Apparently no one thought it cogent to suggest that a wider margin

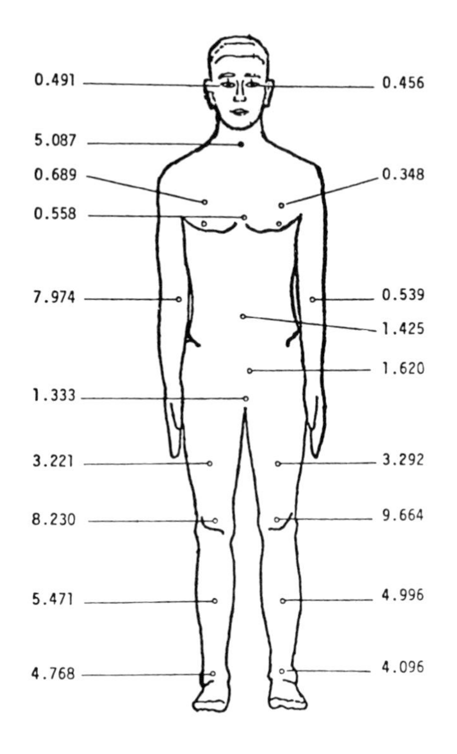

Figure 10
Hotspots—variations in specific absorption rate measure in watts per kilogram. Incident power is $10 mW/cm^2$ giving an average, whole-body SAR of 1.88 W/kg. Reprinted with permission from *IEEE Proceedings* 68(1980): 27. (© 1980 IEEE)

would at least offer greater peace of mind to exposed populations, if not actually offering greater health protection.

Why did C95.IV members so consistently shy away from greater conservatism? Why has policymaking held so long to the upper-limit concept? Since the only possible price of greater conservatism would be possible limitations on the use of RF technology, it seems reasonable to conclude that the main restraint against greater conservatism is a desire to maximize opportunities to expand the use of RF technology. What community has the greatest interest in expanding the use of RF technology? The military and industry, whose values are most strongly represented in C95.1-1982 and other similar policy decisions. At heart C95.1-1982 is a military-industry standard.

This conclusion should come as no surprise. C95 activities are coordinated by the navy and IEEE, two user-oriented organizations. Roughly two of every three C95 members represent military or industrial interests. Many of the scientists who advised during the standard-setting process, including C95.IV chairman Bill Guy, were funded by the military. At every critical juncture the main input into C95.1-1982 came from the user community. That it should as a result reflect the values of that community is natural.

These values contrast sharply with those held by Paul Brodeur and other critics. Critics have consistently argued for a lower-limit approach to standard setting, sometimes to the point of putting potential roadblocks in the way of technological progress. They often have little regard for scientific rigor, accepting, as does Brodeur, any evidence that raises questions, no matter how tenuous it may be. They readily accept anecdotal evidence and argue that more attention should be paid to groups of persons who have been exposed. Above all else they value human health. They worry most about health-related decisions that are made on the basis of partial information. They worry less about slowing technological growth. Most accept the benefits of technology in general but will speak up against technological expansion when they believe that it could affect their lives.

Given these contrasting value perspectives, it is not difficult to understand why there has been an RF debate and why that debate has at times gotten rather heated. If maximum health protection is a dominant value, then the course of past decision making in the United States is inadequate. Reports of injuries could have been thoroughly investigated, exposed populations could have been surveyed in more detail, more chronic, low-level studies could have been funded, standards could have been set more conservatively. If fostering technological expansion is a dominant value, then the criticism that has been leveled at the RF establishment seems grossly unfair. Exposure to moderate and high levels of RF radiation is limited, there is no solid evidence to support claims of widespread injury, some of the subtle effects that have been discovered are of uncertain human consequence, and RF technology does play a critical role in the scheme of modern technological development. There is reason for disagreement and even outrage if the activities of each side are viewed from the perspective of the other's values.

As long as this situation prevails, the chances of ending the microwave debate appear to be slim. Even if the opposing sides eventually agree on the facts, they will likely continue to disagree over interpretations and use. Resolving disagreements over interpretations and use will require an ability to deal with values. If it is true that we will eventually have to confront values, then it would seem to follow that rather than ignoring values as somehow irrelevant ingredients in problem solving, they should be recognized as present and unavoidable. They should be brought into the open and clarified. Persons who study values and their place in society should be consulted with the same frequency and sincerity as scientists. The methods used for integrating values into decision making should be discussed with the same intensity as scientific methods.[12]

Recommendations

Eliminate military influence on RF bioeffects research. Decisions about the use and regulation of RF energy have two fundamental impacts on society: they can influence technological development and they can affect public health. It is possible, although not necessarily certain, that there is conflict between these two impacts. A decision to raise standards could favor technological development, possibly at the expense of health protection. A decision to lower standards could increase public health protection, perhaps at the expense of technological development. Given the potential for conflict, it is essential that persons interested in each aspect of the potential impact of decisions have equal and fair access to decision making.

The most important violation of this principle today can be found in the procedures used to allocate research funds. Significant portions of the support for bioeffects research come from the military, which cannot be viewed as a disinterested party when it comes to making decisions about development versus health. The military has pronounced development interests, which in principle, if not in fact, can influence decisions about health. Thus any work funded by the military can be questioned on the basis of vested interests.

As a major user of RF technology, the military has a right to be concerned about RF bioeffects. The same holds true for industry. Both might reasonably desire or be required (based on a principle of equal distribution of costs and benefits) to make resources available for RF bioeffects research. But neither should have any controlling influence over the way research funds are spent. Decisions relating to the use and control of RF energy are properly public decisions that must be made in public forums where industry, the military, consumers, and anyone else can have equal opportunity to express opinions.

The elimination of military interests from RF bioeffects research would require a change in current funding patterns that could be effected either by reducing military R&D budgets by the amount currently devoted to RF bioeffects research and shiting these funds to agencies that do not have user conflicts or by block allocations from the military to these agencies. However accomplished, the possibility for controlling interests in RF bioeffects research must be eliminated. Only in this way can scientists be freed to get on with their main task—pursuing socially responsible and intellectually challenging investigations—and the public regain its confidence in the bureaucracy that makes decisions about its health.

Exercise congressional responsibility. In the chain of failures that have led to the lack of public policy decisions about the use and regulation of RF energy, the congressional link stands out above all others as the weakest. There is ample reason for singling out Congress as the critical point for change. The only sure input that members of the public have into decision making comes through Congress. Congress sets policy and determines how the nation will be governed. Those who do not agree with policies set by Congress can vote against its members and for other persons.

If Congress does not face up to its responsibilities and instead passes its authority to make decisions on controversial environmental matters to the executive or by default to the courts, public input into decision making is diminished, or lost all together. The public can vote against a representative who favors industrial interests over the public's, or the reverse; it cannot vote against an EPA regulation or a decision handed down by the courts. Loss of input into decision making is bound to result in diminished respect for the decisions that are made.

The consequences of the failure of Congress to take decisive action are many. The vague mandates given to the executive have caused that branch of government to flounder. Local governments have been forced to formulate their own policies to satisfy constituents frustrated by the lack of federal action. Vested interests have looked at indecision as an opportunity to attempt to influence policy. Scientists have had to shoulder the burden of solving many problems that are not scientific ones. In sum the failure of Congress to establish firm guidelines for action has been felt at every level of decision making.

Congress must establish a clear national policy for dealing with the consequences of technological development. The public and industry must know how risks and benefits are to be evaluated so that each can plan accordingly. Official attitudes toward hazards control must be openly debated and eventually agreed on so that judgments can be made about specific projects that touch individual lives. The problems we are facing as a result of technological development are real. So too must be their solutions, in order to allow the latter to be studied, discussed, and, if necessary, changed.

Encourage public involvement. Congressional responsibility probably will not be exercised unless there is active public involvement in the microwave debate. If voters do not ask their representatives to take stands on difficult issues, their representatives will not take stands. If Congress does not ask for changes, it is unlikely that anyone else in government will. Change could always mean the end of jobs and security if decision making were radically restructured. Scientists know that a solution to the RF problem would signal the end of research funds. There is little likelihood that internal pressures within a decision-making establishment will produce change. External pressure, particularly from the public, could.

Public pressure will not affect the microwave debate if it is not grounded on well-informed arguments. Arguments for change that can be easily dismissed do not carry much weight. The public must know what it is talking about when it enters into a debate over the use of RF technology. If it does not, its voice will be quickly and justifiably drowned out by the voices of those who are informed.

Well-informed participation by the public will require better public education. Unfortunately the forces pushing for better public education are not very strong. As long as the public seems satisfied with superficial sensational coverage of controversial issues, the mass media will supply superficial sensational coverage. As long as Congress is permitted to place the burden for making difficult decisions on the executive, it can ignore public dissatisfaction. As long as scientists are more responsible to their funding agencies than to the public, they will be able to dismiss public criticism. The establishment that now oversees policy seemingly has very little to gain from better educating the public and potentially a great deal to lose if viewed within the narrow context of security and the status quo.

If the establishment that controls policy will neither change nor provide the education needed to think critically about change, then the impetus for change and the demand for better education must come from the public. The public must get involved for the purpose of learning more about the problems it wants to address and later as active participants in decision making. In doing so, however, the public must realize that it is taking on a task that can consume considerable amounts of time, money, and energy. It must understand that participation in decision making can require rethinking fundamental values. We live in a world that depends heavily on RF technology for its comforts. It may not be possible to maximize those comforts without infringing on public health. If the public is prepared to accept the challenge posed by this situation fully and responsibly, then it should be encouraged to enter the debate.

Notes

Chapter 1

1. "In the Matter of the Application of Home Box Office, Inc.," transcript of hearings held in Rockaway Township, New Jersey, May 1–October 9, 1980, I:31–32 (hereafter cited as *HBO Hearings*).

2. Ibid., II:81.

3. Ibid.

4. Ibid.

5. Ibid., II:111.

6. Ibid., III:3.

7. Ibid., V:4–5.

8. Ibid., III:92–93.

9. Ibid., III:111.

10. Ibid., XIII:52–53.

11. Ibid., XI:45.

12. Ibid., XI:76.

13. Paul Brodeur, quoted in Michael Matza, "Microwave Menace," *Boston Phoenix*, February 7, 1978.

14. *Consumer Reports* (March 1973): 221.

15. Ibid. (March 1981): 132.

16. Typewritten note (n.d.) by Joe Towne on copies of articles sent to members of the Radar Victims Network, 1976 ff.

17. *FASTA Newsletter* (September 1976).

18. David J. Eisen, director, Research and Information, Newspaper Guild, to me, March 3, 1981.

19. Maurice C. Benewitz, "In the Matter of the Arbitration between Newspaper Guild of New York and the New York Times" (February 1978).

20. *Microwave News* (June 1981): 2.

21. See "Study of RF Sealer Exposures," *Bioelectromagnetics Society Newsletter* (Janury 1984), p. 7, for recent actions taken in Maryland.

Chapter 2

1. *New York Times,* May 23, 1943, sec. 6, p. 14.

2. W. Holzer and E. Weissenberg, *Foundations of Short Wave Therapy* (London, 1935).

3. H. R. Hosmer, "Heating Effects Observed in a High Frequency Static Field," *Science* 68 (1928): 327.

4. C. M. Carpenter and A. B. Page, "The Production of Fever in Man by Short Radio Waves," *Science* 71 (1930): 450–452.

5. W. H. Bell and D. Ferguson, "Effects of Super-High-Frequency Radio Current on Health of Men Exposed under Service Conditions," *Archives of Physical Therapy, X-Ray, Radium* 12 (1931): 477–490.

6. Ibid, pp. 484–485; cf. R. R. Sayers and D. Harrington, "A Preliminary Study of the Physiological Effects of High Temperatures and High Humidities in Metal Mines," *Public Health Reports* 36 (1921): 116.

7. Bell and Ferguson, "Effects," p. 488.

8. L. E. Daily, "A Clinical Study of the Results of Exposure of Laboratory Personnel to Radar and High Frequency Radio," *U.S. Naval Medical Bulletin* 41 (1943): 1032–1056.

9. B. I. Lidman and C. Cohn, "Effect of Radar Emanations on the Hematapoietic System," *Air Surgeon's Bulletin* 2 (1945): 448–449.

10. F. H. Krusen, J. F. Herrick, U. Leden, and K. G. Wakim, "Microkymatotherapy: Preliminary Report of Experimental Studies of the Heating Effect of Microwaves ("Radar") in Living Tissue," *Proceedings of the Staff Meetings of the Mayo Clinic* 22 (1947): 209–224.

11. S. L. Osborne and J. N. Frederick, "Microwave Radiations: Heating of Human and Animal Tissues by Means of High Frequency Current with Wavelength of Twelve Centimeters (The Microtherm)," *Journal of the American Medical Association* 137 (1948): 1036–1040 (hereafter cited as *JAMA*).

12. A. W. Richardson, T. D. Duane, and H. M. Hines, "Experimental Lenticular Opacities Produced by Microwave Irradiations," *Archives of Physical Medicine* 29 (1948): 765–769.

13. C. J. Imig, J. D. Thomson, and H. M. Hines, "Testicular Degeneration as a Result of Microwave Irradiation," *Proceedings of the Society of Experimental Biology and Medicine* 69 (1948): 382–386.

14. J. T. McLaughlin, "A Study of Possible Health Hazards from Exposure to Microwave Radiation," unpublished ms. (Culver City, Calif.: Hughes Aircraft, February 9, 1953). For McLaughlin's comments on the origins of this report, see "Biological Effects of Microwaves," minutes from a Navy Department Conference, Naval Medical Research Institute, Bethesda, Md., April 29, 1953, pp. 5–8 (hereafter cited as Navy Conference [1953]).

15. Navy Conference (1953), pp. 15–16.

16. Ibid., p. 16.

17. "Radar and Cataracts," *JAMA* 150 (1952): 528.

18. F. G. Hirsh and J. T. Parker, "Bilateral Lenticular Opacities Occurring in

a Technician Operating a Microwave Generator," *American Medical Association Archives of Industrial Hygiene* 6 (1952): 512–517.

19. Navy Conference (1953), p. 16.

20. McLaughlin, "Study," p. 26.

21. Private conversations with John Osepchuk, Raytheon Company.

22. A. Hetherington, "Introduction," *Proceedings of the Tri-Service Conference on the Biological Hazards of Microwave Radiation* (AF 18[600]-1180, U.S. Air Force, 1957), p. 1 (hereafter cited as TS-I [1957]).

23. G. M. Knauf, "Chairman's Remarks," *Proceedings of the Fourth Annual Tri-Service Conference* (New York: Plenum Press, 1961), p. 12 (hereafter cited as TS-IV [1960]).

24. S. Michaelson, J. Howland, and R. Dundero, "Review of the Work Conducted at University of Rochester (USAF Sponsored)," in E. G. Parrishall and F. W. Banghart, eds., *Proceedings of the Second Annual Tri-Service Conference on Biological Effects of Microwave Energy,* July 8–10, 1956, RADC-TR-58-54, pp. 175–188 (hereafter cited as TS-II [1958]); and subsequent articles in C. Susskind, ed., *Proceedings of Third Annual Tri-Service Conference on Biological Effects of Microwave Radiating Equipment,* August 25–27, 1959, RADC-TR-59-140 (Berkeley: University of California Printing Department, 1959), pp. 161–238 (hereafter cited as TS-III [1959]).

25. TS-IV (1960), p. 283.

Chapter 3

1. Navy Conference (1953), pp. 17–18.

2. For more information on this process, see Nicholas H. Steneck et al., "The Origins of U.S. Safety Standards for Microwave Radiation," *Science* 208 (June 1980): 1230–1236.

3. Navy Conference (1953), p. 121.

4. Ibid., p. 144.

5. Ibid., p. 44.

6. Ibid., pp. 84–85.

7. Ibid., p. 76.

8. Ibid., pp. 94, 109.

9. Ibid., pp. 86–87.

10. Ibid., p. 102.

11. Ibid., pp. 79, 17.

12. W. B. Deichman and F. H. Stephens, "Microwave Radiation of 10mW/cm^{2} and Factors That Influence Biological Effects at Various Power Densities," *Industrial Medicine and Surgery* 30 (1961): 221–228.

13. The early air force regulations were set out in Urgent Action Technical Orders 31-1-18 (November 1, 1954) and 31-1-511 (June 17, 1957) and RADC Regulation NR 160-1 (May 31, 1957), Rome Air Development Center, Griffiss AFB, New York.

14. TS-II (1958), p. 6; TS-IV (1960), p. 9. Air force sponsorship actually extended back six years, to 1954.

15. H. R. Meahl, "Using Microwaves Safely: 750–30,000 mc," General Electric Technical Information Series, Schenectady, New York, 1956; W. Mumford, "Hazards to Personnel Near High Power UHF Transmitting Antennas," Project Report 717, Bell Laboratories, New York, 1956.

16. TS-II (1958), p. 118.

17. George Knauf, "The Biological Effects of Microwave Radiation on Air Force Personnel," *American Medical Association Archives of Industrial Health* 17 (January 1958): 51.

18. TS-II (1958), p. 118.

19. Ibid., pp. 122–123.

20. Ibid., pp. 120–121.

21. Ibid., p. 120.

22. D. B. Williams and V. J. Nicholson, "Biologic Effects Studies on Microwave Radiation," School of Aviation Medicine, USAF, Randolph AFB, Texas, September 1955, p. 24.

23. TS-II (1958), p. 5.

24. "Report of the General Conference on Standardization in the Field of Radio-Frequency Electromagnetic Radiation Hazards," American Standards Association, XX 4772, May 4, 1959.

25. J. J. Dunn, chief, Standards Division, Department of Defense, memorandum, December 2, 1958.

26. A. L. Van Emden to J. J. Anderson, secretary, American Institute of Electrical Engineers, June 5, 1959.

27. A. L. Van Emden, Department of the Navy, Bureau of Ships, memorandum, Code 621B, December 8, 1959.

28. Ibid.

29. J. Paul Jordan to A. L. Van Emden, January 22, 1960.

30. American Standards Association (hereafter cited as ASA) Sectional Committee C95, minutes, February 15, 1960.

31. S. David Hoffman, "Joint Sponsors of ASA Sectional Committee on Radiation Hazards, C95," memorandum, April 12, 1960; and Travel Report 9670 submitted by Glenn Heimer, "Subject: Visit to University of Pennsylvania, Philadelphia, Pennsylvania," May 3, 1960. AIEE had become the advisory cosponsor by the second C95 meeting (April 24, 1962) and the navy the primary cosponsor.

32. C. M. Cormack, Jr., Bureau of Naval Weapons, to Major General E. P. Mechling, American Ordnance Association, March 28, 1960.

33. ASA Sectional Subcommittee C95.4, minutes, January 18, 1961.

34. Glenn Heimer, secretary ASA C95, to Herman Schwan, September 5, 1962.

35. ASA Sectional Committee C95, minutes, November 20, 1962.

36. Ibid., p. 5.

37. ASA Sectional Subcommittee C95, "First Draft, Interim Standard: Safety Level and/or Tolerances with Respect to Personnel," June 24, 1963.

38. ASA Sectional Subcommittee C95, minutes, February 19, 1963, pp. 3–4.

39. USA Standards Institute, USA Standard C95.1-1966, "Safety Level of Electromagnetic Radiation with Respect to Personnel," November 1966.

40. Herman Schwan to Sol Michaelson, November 10, 1970.

41. USA Standards Institute, "Personnel of ASA Sectional Committee C95, Radio-Frequency Radiation Hazards," April 1, 1965.

42. Herman Schwan to John Anderson, June 11, 1965.

43. USA Standards Institute, "Tally Sheet" for "Certification of Letter Ballot of the Electrical Standards Board on . . . C95.1 and C95.2," October 20, 1966.

44. Robert Ullman, U.S. Army Electronics Command, Fort Monmouth, New Jersey, to Commander, Naval Ship Engineering Center, memorandum, November 3, 1966.

45. Glenn Heimer to John E. Gerling, president, International Microwave Institute, c. November–December 1966.

46. The earliest accurate report on the Soviet standard that I have been able to find is John J. Turner, "The Effects of Radar on the Human Body (Results of Russian Studies on the Subject)," U.S. Army Ordnance Missile Command, Bell Telephone Laboratories, Whippany, New Jersey (July 1962), pp. 1–9.

47. Sol Michaelson, "Standards for Protection of Personnel against Nonionizing Radiation," *American Industrial Hygiene Association Journal* 35 (December 1974): 781.

48. Ibid.

49. Karel Marha, "Maximum Admissible Values of HF and UHF Electromagnetic Radiation at Work Places in Czechoslovakia," in Stephen F. Cleary, ed., *Biological Effects and Health Implications of Microwave Radiation,* proceedings from a symposium held in Richmond, VA., September 17–19, 1969 (Washington, D.C., 1970), pp. 188–191 (hereafter cited as Richmond Symposium [1969]).

50. Ibid.

51. S. Baranski and P. Czerski, *Biological Effects of Microwaves* (Stroudsburg, Pa.: Dowden, Hutchinson, and Ross, 1976), p. 154.

52. This method is discussed by Marha, Richmond Symposium (1969), pp. 183–191, and "Microwave Radiation Safety Standards in Eastern Europe," *IEEE Transactions on Microwave Theory and Techniques* 19 (1971): 165–168; and Baranski and Czerski, *Biological Effects,* pp. 170–187.

53. N. H. Steneck, "Subjectivity in Standards: The Case of ANSI C95.1-1982," *Microwaves and RF* (May 1983), pp. 137ff.

Chapter 4

1. J. W. Schereschewsky, "The Physiological Effects of Currents of Very High Frequency (135,000,000 to 8,300,000 Cycles per Second)," *Public Health Reports* 41 (September 1926): 1939–1963.

2. J. W. Schereschewsky, "The Action of Currents of Very High Frequency upon Tissue Cells: A. Upon a Transplantable Mouse Sarcoma," *Public Health Reports* 43 (April 1928): 927–945.

3. R. V. Christie, "An Experimental Study of Diathermy: VI. Conduction of

High Frequency Currents through the Living Cells," *Journal of Experimental Medicine* 48 (1928): 235–246.

4. R. V. Christie and A. L. Loomis, "Relationship of Frequency to Physiological Effects of Ultra-High Frequency Currents," *Journal of Experimental Medicine* 49 (1929): 303–321.

5. Ibid., p. 320.

6. J. W. Schereschewsky, "Biological Effects of Very High Frequency Electromagnetic Radiation," *Radiology* 20 (1932): 246–253.

7. R. R. Mellon, W. T. Szymanowski, and R. A. Hicks, "An Effect of Short Electric Waves on Diphtheria Toxin Independent of the Heat Factor," *Science* 72 (1930): 174–175.

8. R. A. Hicks and W. T. Szymanowski, "The Biologic Action of Ultrahigh Frequency Currents: Further Studies," *Journal of Infectious Diseases* 50 (1932): 466–472.

9. Among the many books written on short-wave therapy, see E. Schliephake, *Kurzwellen therapie* (1932), trans. K. Brown, *Short Wave Therapy* (London, 1935); and W. Bierman and M. M. Schwarzchild, *The Medical Applications of Short-wave Current* (Baltimore, 1938).

10. F. H. Krusen, "Short Wave Diathermy, A Preliminary Report," *JAMA* 104 (1935): 1237–1239.

11. B. Mortimer and S. L. Osborne, "Tissue Heating by Short Wave Diathermy," *JAMA* 104 (1935): 1413–1419.

12. Ibid., p. 1418.

13. Ibid., p. 1419.

14. H. F. Wolf, "The Physiological Basis of Short Wave Therapy," *JAMA* 104 (1935): 76.

15. S. M. Horvath, R. N. Miller, and B. K. Huff, "Heating of Human Tissues by Micro Wave Radiation," *American Journal of the Medical Sciences* 216 (October 1948): 430–436; J. W. Gersten, K. G. Wakim, J. F. Herrick, and F. H. Krusen, "The Effect of Microwave Diathermy on the Peripheral Circulation and on Tissue Temperature in Man," *Archives of Physical Medicine* 30 (January 1949): 7–25.

16. W. W. Salisbury, J. W. Clark, and H. M. Hines, "Physiological Damages due to Microwaves," unpublished report (Cedar Rapids, Iowa: Collins Radio Co., December 27, 1948); A. W. Richardson, T. D. Duane, and H. M. Hines, "Experimental Lenticular Opacities Produced by Microwave Irradiations," *Archives of Physical Medicine* 28 (December 1948): 765–769.

17. C. J. Imig, J. D. Thomson, and H. M. Hines, "Testicular Degeneration as a Result of Microwave Irradiation," *Proceedings of the Society for Experimental Biology and Medicine* 69 (1948): 382–386.

18. J. E. Boysen, "The Biological Effects of Electro Magnetic Radiation (Hertzian Waves)" (Ph.D. diss., University of Cincinnati, 1951), p. 8.

19. H. M. Hines and J. E. Randall, "Possible Industrial Hazards in the Use of Microwave Radiation," *Electrical Engineering* 71 (October 1952): 879.

20. D. B. Williams and W. J. Nicholson, "Biologic Effects Studies on Microwave Radiation: An Appraisal of the Biological Effects of Current USAF 'S'

Band, Ground Radar Transmitters," School of Aviation Medicine, Randolph AFB, Texas, 1955.

21. Ibid., pp. 12, 24. The original study found that infrared damage occurred only if above normal temperatures were induced. The effects seen at the higher temperatures induced by infrared radiation were the same as those seen at lower temperatures with microwave irradiation.

22. "Summary of Results of UHF Radiation Hazard Experiments at Lincoln Laboratory, MIT," TS-I (1957), pp. 104–108, and S. J. Fricker, moderator of "Microwave Exposure Discussion," TS-I (1957), pp. 79–80.

23. See the comments of F. W. Hartman, TS-I (1957), p. 81.

24. S. Gunn, T. Gould, and W. A. D. Anderson, "The Effects of Microwave Radiation (24,000 mc) on the Male Endocrine System of the Rat," TS-IV (1960), pp. 99–115, partially duplicated earlier studies of testicular effects and suggested further work. Duplicate or follow-up studies were not attempted.

25. B. L. Vosburgh, "Recommended Tolerance Levels of M-W Energy: Current Views of the General Electric Company's Health and Hygiene Service," TS-II (1958), p. 123.

26. Williams and Nicholson, "Biologic Effects Studies," p. 17; C. I. Barron, A. A. Love, and A. A. Baraff, "Physical Evaluation of Personnel Exposed to Microwave Emanations," *Aviation Medicine* 26 (December 1955): 447.

27. R. Tallarico and J. Ketchum, "Certain Pilot Studies Were Conducted by Dr. Robert Tallarico and John Ketchum of the Department of Physiology, University of Miami," presented by W. B. Deichmann, in TS-III (1959), pp. 75–76.

28. S. A. Bach, M. Baldwin, and S. Lewis, "Some Effects of Ultra-High Frequency Energy on Primate Cerebral Activity," TS-III (1959), p. 83.

29. "High Intensity Radiation Produces Convulsions, Death in Monkey," *Aviation Week* 70 (May 1959): 29–30.

30. D. E. Goldman, "Comments on the Paper Titled 'Neurological Effects of Radio-Frequency Energy,' " by S. A. Bach, S. A. Lewis, and M. Baldwin, TS-III (1959), p. 93.

31. R. D. McAfee, "Neurophysiological Effect of 3-cm Microwave Radiation," *American Journal of Physiology* 200 (1961): 194; cf. TS-IV (1960), pp. 251–260.

32. A. H. Frey, "Auditory System Response to Radio Frequency Energy," *Aerospace Medicine* 32 (December 1961): 1140–1142.

33. TS-IV (1960), pp. 10–11.

34. Z. V. Gordon et al., "Biological Effects of Microwaves of Small Intensity," and A. N. Obrosov and V. G. Jasnogorodski, "A New Method of Physical Therapy—Pulsed Electric Field of Ultra-High Frequency," in *Digest of the 1961 International Conference on Medical Electronics,* ed. P. L. Frommer (New York, The Conference Committee, 1961), pp. 153, 155–156.

35. Y. A. Kholodov, "The Effect of an Electromagnetic Field on the Central Nervous System," trans. Translation Services Branch, Foreign Technology Division, Wright Patterson Air Force Base, Ohio, August 9, 1962, FTD-TT-62-1107/1, AD 284123.

36. A. S. Pressman and N. A. Levitina, "Nonthermal Action of Microwaves on the Rhythm of Cardiac Contractions in Animals, Report II: Investigation of Pulsed Microwave Action," trans. Translation Services Branch, Foreign Technology Division, Wright Patterson Air Force Base, Ohio, June 21, 1962, FTD-TT-62-501/1 + 2, AD 283882.

37. J. J. Turner, "The Effects of Radar on the Human Body (Results of Russian Studies on the Subject)," U.S. Army Ordnance Missile Command, Liaison Office, Bell Telephone Laboratories, Whippany, New Jersey, July 16, 1962, AD 278172. Turner's report is based on a translation of a 1960 article by A. A. Letavet and D. Z. Gordon, "The Biological Action of Ultra High Frequencies."

38. This and all subsequent information on the UCLA meeting is taken from the unpublished minutes: "Neurological Responses to External Electromagnetic Energy (A Critique of Currently Available Data and Hypotheses)," cosponsored by the Brain Research Institute, UCLA, and the Air Force Systems Command, July 11, 1963, USAF Contract AF 18(600)-2057.

39. Ibid., pp. 4–5, 48–51.

40. Ibid., pp. 9–21.

41. Ibid., pp. 51–57.

42. Ibid., pp. 80–94.

43. John T. McLaughlin, "Tissue Destruction and Death from Microwave Radiation (Radar)," *California Medicine* 86 (1957): 336–339.

Chapter 5

1. The major sources of information made public are the nine bound volumes released by the Department of State as Pandora Project (hereafter cited as USSD-PP) and "Microwave US-USSR," vols. I–VIII (hereafter cited as USSD-MW).

2. Information on the history of the signal is contained in R. C. Mallalieu, "A Model of the Microwave Intensity Distribution within the U.S. Embassy in Moscow, 1966–1977," Fleet Systems Department, Applied Physics Laboratory, Johns Hopkins University, August 1980, FS-80-166.

3. This memorandum, which was declassified on May 20, 1981, was not contained in the original Pandora materials made public in 1977. My copy was kindly supplied by its author.

4. Personal correspondence from Dean Rusk, March 19, 1982.

5. Department of State, memorandum, G. M. Gentile, deputy assistant secretary for security, Department of Sate, to Dr. Lewis K. Woodward, May 17, 1965; USSD-MW I, no. 21; and B. W. Augenstein, Office of the Director of Defense Research and Engineering, to Drs. Brown and Fubine, memorandum, May 13, 1965, p. 1. The involvement of the White House is mentioned in "Memorandum for the Director, Advanced Research Projects Agency, Subject: Justification Memorandum for Project PANDORA," October 15, 1965.

6. Cited in n. 5 above. This memorandum is not contained in the Pandora file and was made available to me by ARPA in response to a general inquiry for additional information.

7. Ibid., p. 3.

8. USSD-MW I, nos. 17, 18, 19, 21.

9. M. Zaret, S. Cleary, B. Pasternack, and M. Eisenbud, "Occurrence of Lenticular Imperfections in the Eyes of Microwave Workers and Their Association with Environmental Factors," Institute of Industrial Medicine, New York University-Bellevue Medical Center, New York 1961, RADC-TN-61-226, AD-266-831.

10. B. L. Herndon, M. A. Giagle, and J. J. Downs, "Biological Effects of Microwave Radiation," for presentation at the AIBS meetings Boulder, Colorado, August 26, 1964, USSD-MW I, no. 16. Although the results of this study were not published, a copy of the oral presentation did make its way into the State Department "Moscow Medical" file, so presumably it was used to make decisions.

11. Dr. G. Mishtowt, assistant medical director, Department of State, to Dr. R. Stevenson, head, Eye Clinic, National Naval Medical Center, April 5, 1966, USSD-MW I, no. 18.

12. R. O. Ranheim to Dr. C. Klontz, deputy medical director, Department of State, April 13, June 15, 1967, USSD-MW I, nos. 36, 41.

13. C. Klontz to C. Weiss, memorandum, [January 9, 1967], USSD-MW I, no. 55.

14. Ibid., p. 3.

15. For the initiating correspondence and initial contract, see USSD-MW I, no. 25.

16. G. Mishtowt to R. Ranheim, June 21, 1968, USSD-MW I, no. 55.

17. C. Nydell, deputy medical director, Department of State to C. Jacobson, June 25, 1969; and C. Jacobson, "Final Report," contracts SCC 31252, 31732, 31759, 31759 Mod. 1, 31759 Mod. 2, received by the Department of State, August 4, 1969.

18. Paul Brodeur, *The Zapping of America* (New York: W. W. Norton, 1977), p. 133.

19. Department of State, memorandum for "limited official use," March 20, 1969, summarizing the outside reviews of Jacobson's work.

20. Details of the testing program are contained in Jacobson's reports to the Department of State, which were released in response to Freedom of Information Act requests from Associated Press reporter Barton Reppert.

21. These recommendations appear in a progress report on SCC 31732, dated November 15, 1968, and were repeated again in a similar report dated February 4, 1969.

22. Jacobson, "Final Report," p. 4.

23. USSD-MW I, no. 29.

24. Pandora, minutes, April 4, 1969, p. 4, USSD-PP, no. 8.

25. D. A. Hungerford to G. I. Mishtowt, January 14, 1969.

26. J. H. Tjio to H. Pollack, memorandum, May 29, 1978.

27. "Visit with Dr. McKusick, Johns Hopkins University," report, February 20, 1969; V. A. McKusick to C. C. Nydell, March 7, 1969.

28. Department of State memorandum, March 20, 1969.

29. C. C. Nydell to J. H. Tjio, March 17, 1969.

30. Hungerford reviewed slides 1–50, 70–76, 79, 84, 101, 102, Tjio reviewed slides 94–100, and Hirschhorn reviewed slides 53, 61, 63, 66, 68, 83, 103, 106, 107, 110, 111, and 112.

31. K. Hirschhorn to C. C. Nydell, June 18, 1969.

32. Pandora, minutes, April 4, 1969, USSD-PP, no. 8.

33. Gresinger's comments were made in interviews with Associated Press reporter Barton Reppert and have been widely reported since.

34. The progress of these studies can be traced in USSD-MW I, nos. 19, 20.

35. USSD-MW I, nos. 56, 58.

36. G. H. Heilmeier to D. Johnson, February 5, 1976, USSD-MW-PP, no. 1.

37. G. H. Heilmeier to Senator W. G. Magnuson, September 15, 1977, in U.S. Senate, Committee on Commerce, Science and Transportation, *Radiation Health and Safety, Hearings on Oversight of Radiation Health and Safety,* 95th Congress, 1st sess., 1977, p. 1184 (hereafter cited as Senate Hearings [1977]).

38. J. R. Hamer, "Biological Entrainment of the Human Brain by Low Frequency Radiation," unpublished paper, August 1965. The copy used in preparing this book came from the ARPA files.

39. R. S. Cesaro, "Program Plan No. 562, Pandora," Advanced Sensors Program, ARPA, October 15, 1965. Any documents cited that have not been published in Senate Hearings (1977) were received directly from ARPA.

40. R. S. Cesaro, memorandum, "Justification Memorandum for Pandora," ARPA, October 15, 1965.

41. E. V. Byron, "Operational Procedure for Project Pandora Microwave Test Facility," Johns Hopkins Applied Physics Laboratory, October 1966; Johns Hopkins Applied Physics Laboratory, "Project Pandora (U): Final Report," November 1966.

42. Details of the initial test results are contained in two memos prepared by R. S. Cesaro for the director of ARPA, both titled "Project Pandora—Initial Test Results." The first is dated December 15, 1966, the second December 20, 1966.

43. Cesaro, "Initial Test Results," December 15, 1966.

44. Cesaro, "Initial Test Results," December 20, 1966, pp. 2–3.

45. Ibid.

46. Undated, unsigned memo, c. November 1966, USSD-MW I, no. 52. The earliest report of the lower measurements is probably the CIA Memorandum for Deputy Director, R&D, [ARPA], dated September 13, 1967, cited by R. S. Cesaro, "Memorandum for Director, Defense Research & Engineering, Subject: Project Bizarre," September 27, 1967.

47. Cesaro, memorandum, September 27, 1967, p. 3.

48. Institute of Defense Analysis Review Panel, "Flash Report of Pandora/ Bizarre Briefing," January 14, 1969, USSD-PP, no. 6.

49. For a description of the literature search, see Janet Healer, Allied Research Associates, to Director, ARPA, Document No. 9G61-2, February 15, 1969.

50. J. F. Kubis, "The Saratoga Study," unpublished paper, May 8, 1969, esp. p. 5.

51. Pandora, minutes, August 12–13, 1969, p. 9, USSD-PP, no. 12; cf. ibid., April 12, 1969, p. 1, USSD-PP, no. 8.

52. IDA Review Panel, "Flash Report."

53. Pandora, minutes, April 21, 1969, p. 7.

54. Ibid., June 18, 1969, p. 2.

55. Ibid., August 12–13, 1969, p. 10.

56. Joseph F. Kubis, "On the Evaluation of Data Associated with Pandora (Preliminary Report)," ms., December 4, 1969.

57. Brodeur, *Zapping of America.*

58. Information on nonbiological testing that followed the discovery of the Moscow signal is still classified.

59. "Agreement Transfer of Project Pandora," effective date, July 1, 1970, unnumbered ARPA papers.

60. USSD-MW I, no. 48, p. 2.

61. Pandora, minutes, April 21, 1969, p. 5.

62. Ibid., June 18, 1969, p. 2.

63. Ibid., January 12, 1970, p. 2.

Chapter 6

1. *New York Times,* May 9, 1954.

2. Ibid., August 21, 1960.

3. U.S. House, Committee on Interstate and Foreign Commerce, Subcommittee on Public Health and Welfare, *Electronic Products Radiation Control, Hearings on H.R. 10790,* 90th Congress, 1st sess., 1967, p. 2 (hereafter cited as House Hearings [1967]).

4. IEEE, Joint Technical Advisory Committee, *Spectrum Engineering, The Key to Progess* (New York: IEEE, 1968), p. 84.

5. Ibid., p. 6.

6. U.S. Senate, Committee on Commerce, *Radiation Control for Health and Safety Act of 1967, Hearings on S. 2067, S. 3211, and H.R. 10790,* 90th Congress, 1st, 2nd sess., 1967–1968, p. 2 (hereafter cited as Senate Hearings [1967–1968]).

7. Ibid., pp. 407–408.

8. Ibid., pp. 709–717.

9. Ibid., p. 953.

10. Ibid., p. 762.

11. Ibid., pp. 65–66.

12. *New York Times,* January 18, 1968.

13. Public Law 90-602, October 18, 1968.

14. Ibid.

15. For brief descriptions of these and other regulatory activities in the ex-

ecutive, see L. David, "A Study of Federal Microwave Standards," report prepared for the U.S. Department of Energy, Office of Energy Research, DOE/ER/10041-02 (August 1980).

16. Senate Hearings (1967–1968), pp. 788–789.

17. *New York Times,* May 24, 1969, p. 33, January 5, 1970, p. 26. The January report was based on Bureau of Radiological Health, HEW, *Microwave Oven Surveys,* Report DEP 69-7 (December 1969).

18. P. A. Breysse, "Microwave Uses on Campus: A Study of Environmental Hazards," *Journal of Microwave Power* 4 (1969): 25–28.

19. Ibid., p. 27.

20. J. M. Osepchuk, "A Review of Microwave Oven Safety," *Journal of Microwave Power* 13 (1978): 13.

21. Ibid., p. 14.

22. *Code of Federal Regulations,* title 21, chapter I, subchapter J, paragraph 1030.10C1, October 6, 1971.

23. For relevant BRH publications, see nos. 72-8012, 72-8015, 75-8018, 76-8004, 79-8030, 80-8106. See also U.S. House, Committee on Science and Technology, Subcommittee on Investigations and Oversight, *Potential Health Effects of Video Display Terminals and Radio Frequency Heaters and Sealers, Hearings,* 97th Congress, 1st sess., 1981 (hereafter cited as House Hearings [1981]).

24. *Federal Register,* June 9, 1975, p. 24579.

25. Occupational Safety and Health Administration, *Occupational Safety and Health Decisions* (1975–1976), no. 20,379.

26. For a review of the RF sealer problem, see House Hearings (1981), pp. 456–522.

27. Senate Hearings (1977), p. 577. See also *Occupational Safety and Health Decisions* (1977), no. 21,656.

28. D. L. Conover, W. H. Parr, E. L. Sensinfaffer, and W. E. Murray, Jr., "Measurement of Electric and Magnetic Field Strengths from Industrial Radiofrequency (15–40.68 MHz) Power Sources," in *Biological Effects of Electromagnetic Waves: Selected Papers of the USNC/URSI Annual Meeting* (Washington, D.C.: BRH, 1976), 2:356–362.

29. Senate Hearings (1977), p. 578.

30. "RF/Microwave Criteria Document: External Review Draft," 2 vols., circulated in ms. form in 1979 and 1980; see also Z. R. Glaser, "Basis for the NIOSH Radiofrequency and Microwave Radiation Criteria Document," in *Nonionizing Radiation—Proceedings of an ACGIH Topical Symposium,* November 26–28, 1979 (Washington: ACGIH, 1980), pp. 103–116.

31. U.S. House, Committee on Science and Technology, Subcommittee on Natural Resources and Environment, *Research on Health Effects on Nonionizing Radiation, Hearing,* 96th Congress, 1st sess., 1979, p. 86 (hereafter cited as House Hearings [1979]).

32. D. E. Janes, Jr., "Population Exposure to Radiowave Environments in the United States," in *Life Cycle Problems and Environmental Technology,* Proceedings of the Twenty-sixth Annual Technical Meeting of the Institute of En-

vironmental Sciences, Philadelphia, May 12–14, 1980 (Mount Prospect, Ill.: Institute of Environmental Sciences, 1980), pp. 154–158; R. A. Tell and E. D. Mantiply, "Population Exposure to VHF and UHF Broadcast Radiation in the United States," *Proceedings of the IEEE* 68 (January 1980): 6–12.

33. Senate Hearings (1977), p. 82.

34. House Hearings (1979), p. 196.

35. *Federal Register,* April 27, 1981, p. 23696.

36. Senate Hearings (1967), p. 665.

37. FCC, *Notice of Inquiry,* General Docket no. 79-144. For a summary of the responses, see R. F. Cleveland, "Results of the Federal Communications Commissions Notice of Inquiry on RF and Microwave Bioeffects," in N. H. Steneck, ed., *Risk/Benefit Analysis: The Microwave Case* (San Francisco: San Francisco Press, 1982), pp. 69–96.

38. Karen Massey, "The Challenge of Nonionizing Radiation: A Proposal for Legislation," *Duke Law Journal* 105 (1979): 105–188.

39. House Hearings (1979), p. 107.

40. Ibid., p. 35.

41. *Federal Register,* December 23, 1982, p. 57338.

42. *Microwave News* (January–February 1982): 5.

43. House Hearings (1979), p. 171.

44. Executive Order 12046, March 27, 1978.

45. *Microwave News* (July–August 1981): 1–2; ibid. (January–February 1983): 2.

46. House Hearings (1979), p. 117.

47. Ibid., p. 202.

48. *Microwave News* (June 1981): 2.

49. *Solar Power Satellite Research, Development and Evaluation Program Act of 1979,* H.R. 2335, February 22, 1979.

50. U.S. House, Committee on Science and Technology, Subcommittee on Space Science and Applications, *Solar Power Satellite, Hearings,* 96th Congress, 1st sess., 1979, p. 11.

51. *Solar Power Satellite Research, Development and Evaluation Program Act of 1981,* H.R. 497, January 5, 1981.

Chapter 7

1. *Richmond Symposium* (1969), p. 6.

2. Ibid., p. 3.

3. *Biological Effects of Microwaves: Future Research Directions,* panel discussion chaired by Lt. Col. Alvin M. Burner, USAF, March 22, 1968 (San Francisco: San Francisco Press, 1968).

4. Ibid., p. 19.

5. Ibid., p. 8.

6. Ibid., pp. 5–6.

7. Ibid., p. 20.

8. Office of Telecommunications Management, Executive Office of the President, minutes, "First Meeting of the Electromagnetic Radiation Management Advisory Council," March 27, 1979, p. 5 (hereafter referred to as ERMAC, minutes). See also ERMAC, minutes, June 5, 1969, p. 5, July 9, 1969, p. 3, September 16, 1969, p. 2. The program was circulated in manuscript as "a recommendation by the Electromagnetic Radiation Management Advisory Committee to the Director of Telecommunications Policy" (December 1971) (hereafter cited as ERMAC, *Program*). For a later published version, see "A Technical Review of the Biological Effects of Non-ionizing Radiation," report prepared for the Office of Science and Technology Policy by an ad hoc working group, May 15, 1978.

9. ERMAC, *Program,* pp. 2–3.

10. *Richmond Symposium* (1969), p. 256.

11. Ibid., p. 254.

12. Ibid., p. 235.

13. Herman Schwan to Glenn Heimer, secretary, ANSI C95, November 4, 1965; and ANSI C95, minutes, October 22, 1969, p. 3. Schwan was out of the country the year before and did not take an active role in organizing C95.IV affairs after late 1967; see ANSI C95, minutes, June 24, 1968.

14. Saul Rosenthall to O. M. Salate, colleague of Herman Schwan, August 8, 1968.

15. Schwan outlined his views on C95.1-1966 in a letter to Sol Michaelson dated November 10, 1970.

16. A clear summary of these issues can be found in a letter from Rosenthall, June 17, 1969, convening a C95.IV meeting on July 2, 1969.

17. Saul Rosenthall to John Villforth, October 14, 1969. This letter was discussed but not endorsed at an ANSI C95 meeting, October 22, 1969. See ANSI C95, minutes, October 22, 1969, p. 2.

18. Arthur W. Guy to ANSI C95.IV members, July 10, 1970. The study group dealing with frequency effects was later divided into two committees, covering effects above and below 30 MHz, respectively.

19. Milton Zaret to ANSI C95 members, April 15, 1970; "Recommendations for modification of Standard . . . (USAS-C95.1-1966)."

20. ANSI C95.IV, minutes, November 17, 1970, pp. 1–2.

21. Ibid., p. 1.

22. "Research Needed for Setting of Realistic Safety Standards," report by ANSI C-95 Subcommittee IV on Safety Levels and/or Tolerances with Respect to Personnel, A. W. Guy, chairman, approved June 26, 1972, and published August 1, 1972. Similar priorities are given in C. C. Johnson, "Research Needs for Establishing a Radio Frequency Electromagnetic Radiation Safety Standard," *Journal of Microwave Power* 8 (1973): 367–388.

23. John T. McLaughlin, "Tissue Destruction and Death from Microwave Radiation (Radar)," *California Medicine* 86 (1957): 336–339.

24. J. T. McLaughlin, "Health Hazards from Microwave Radiation," *Western Medicine* 3 (1962): 126–132.

25. "Request for Review Article," prepared by William M. Silliphant, direc-

tor, Armed Forces Institute of Pathology, for CO, U.S. Army Environmental Health Laboratory, Army Chemical Center, MD, MEDEM-SD, July 8, 1957.

26. Senate Hearings (1967), p. 711.

27. *JAMA* 216 (1971): 1651.

28. *Ibid.* 217 (1971): 1394.

29. U.S. Senate, Committee on Commerce, *Radiation Control for Health and Safety, Hearings on Public Law 90-602,* 93rd Congress, 1st sess., 1973, p. 136 (hereafter cited as Senate Hearings [1973]).

30. Pathology Laboratory, Culver City Hospital, report, specimen number CA-4-5-54, patient Stanley Ries 35623, performed April 5, 1954, by Milton Rosenthal, MD.

31. Herman Schwan et al., "Hazards Due to Total Body Irradiation by Radar," *Proceedings of the Institute of Radio Engineers* 44 (1956): 1572–1581.

32. Charlotte Silverman, "Epidemiologic Approach to the Study of Microwave Effects," *Bulletin of the New York Academy of Medicine* 55 (December 1979): 1166–1181.

33. Ibid.; M. M. Zaret, S. F. Cleary, and B. S. Pasternack, *A Study of Lenticular Imperfections in the Eyes of a Sample of Microwave Workers and a Control Population,* Contract Reports for Rome Air Development Center, RADC-TDR-63-125, March 15, 1963; B. H. Cohen et al., "Parental Factors in Down's Syndrome—Results of the Second Baltimore Case-control Study," in *Population Cytogenetics: Studies in Humans,* ed. E. B. Hook and I. H. Porter (New York: Academic Press, 1977), pp. 301–352; J. A. Burdeshaw and S. Schaffer, *Factors Associated with the Incidence of Congenital Anomalies: A Localized Investigation,* Final Report, Environmental Protection Agency, Contract 68-02-0791, March 31, 1976; C. D. Robinette, C. Silverman, and S. Jablon, "Effects upon Health of Occupational Exposure to Microwave Radiation (Radar)," *American Journal of Epidemiology* 112 (1980): 39–53; A. M. Lillienfeld et al., *Foreign Service Health Status Study,* Final Report to Department of State, Contract No. 6025-619073, July 13, 1978 (hereafter cited as *FSHSS*).

34. *FSHSS,* pp. 93, 246–248.

35. *Microwave News* (January 1981): 1, 7; ibid. (July–August 1981): 3.

36. H. Pollack, "Epidemiologic Data on American Personnel in the Moscow Embassy," *Bulletin of the New York Academy of Medicine* 55 (December 1979): 1186; cf. *FSHSS,* p. 246.

37. Battelle, Human Affairs Research Center, "Final Environmental Impact Statement, RCA Earth Station Bainbridge Island, Washington," prepared for Kitsap County, Washington (June 1982), p. 30.

38. Ibid., pp. 47, 42.

39. Ibid., pp. F-10, F-4, F-11.

40. Ibid., pp. 65–162.

41. R. Carpenter, "The Action of Microwave Radiation on the Eye," *Journal of Microwave Power* 3 (1968): 3–20; "Experimental Microwave Cataract: A Review," *Richmond Symposium* (1969): 76–81.

42. S. Michaelson, "Human Exposure to Nonionizing Radiant Energy—Potential Hazards and Safety Standards," *Proceedings of the IEEE* 60 (1972): 410.

43. Senate Hearings (1973), p. 144.

44. M. Zaret and W. Snyder, "Cataracts and Avionic Radiations," *British Journal of Ophthalmology* 61 (1977): 383.

45. See the comment and reply in Michaelson, "Human Exposure," pp. 1237–1238.

46. Milton Zaret, "Cataracts Following Use of Microwave Oven," *New York State Journal of Medicine* (October 1974): 2032–2048.

47. Ibid., p. 2037.

48. Ibid., p. 2038.

49. Ibid., pp. 2032, 2034.

50. Ibid., p. 2035.

51. Ibid., p. 2033, 2036, 2037.

52. Budd Appleton et al., "Microwave Lens Effects in Humans," *Archives of Ophthalmology* 88 (1972): 259–262, and 93 (1975): 257–258.

53. See "Deposition of Milton M. Zaret, in the case of Mulhauser and Ryan versus Litton Systems, Inc.," October 2, 1975.

54. Battelle, "EIS," p. 31.

55. U.S. Air Force School of Aerospace Medicine, Brooks Air Force Base, Texas, contract F 33615-78-C-0631, September 22, 1978–May 31, 1980.

56. U.S. Air Force School of Aerospace Medicine, Brooks Air Force Base, Texas, contract F 33615-80-C-0612, September 1, 1980–February 28, 1982. For a description of the experimental procedures used, see "Effects of Long-Term Low-Level RFR Exposure on Rats," Technical Report, March 1982, prepared for USAF School of Aerospace Medicine, Brooks Air Force Base, Texas, F 33615-80-C-0612.

57. A. W. Guy, C.-K. Chou, R. B. Johnson, and L. L. Kunz, "Study of Effects of Long-term Low-level RF Exposure on Rats: A Plan," *Proceedings of the IEEE* 68 (1980): 92–97.

58. Allan Frey to Robert Frazier, June 9, 1980.

59. Headquarters Aeronautical Systems Division, USAF, Wright-Patterson Air Force Base, Ohio, "Request for Proposal F33615-80-R-0612," February 14, 1980.

60. See "Effects of Long-Term Low-Level RFR Exposures," pp. 192–194.

61. Rochelle Medici, "Where Has All the Science Gone," in Steneck, *Risk/Benefit Analysis,* p. 184.

62. Battelle, "EIS," p. 124. This response was originally intended for publication in Steneck, *Risk/Benefit Analysis* but later was withdrawn at Guy's request; Steneck, *Risk/Benefit Analysis,* p. 226.

63. Battelle, "EIS," p. 124.

64. Shandala, for example, tested animals on a more intense time schedule and discovered as well that effects tended to disappear during long-term exposure; see M. G. Shandala et al., "Study of Nonionizing Microwave Radiation Effects upon the Central Nervous System and Behavior Reactions," *Environmental Health Perspectives* 30 (1979): 115–121. Guy's six-week test cycle would then not necessarily detect effects discovered by Shandala.

65. For summaries of yearly funding figures, see the reports on *Program for Control of Electromagnetic Pollution of the Environment,* issued by NTIA, March 1973, May 1974, April 1975, June 1976, March 1979.

66. A. H. Frey et al., "Neural Function and Behavior: Defining the Relationship," *Annals of the New York Academy of Sciences* 247 (1975): 433–438; E. N. Albert, "Reversibility of Microwave-Induced Blood-Brain-Barrier Permeability," *Radio Science* 14 (1979): 323–327; K. J. Oscar and T. D. Hawkins, "Microwave Alternations of the Blood-Brain-Barrier System of Rats," *Brain Research* 126 (1977): 281–293.

67. J. H. Merritt et al., "Some Biologic Effects of Microwave Energy Directed Toward the Head," unpublished and unclassified abstract circulated to participants. The conference was sponsored by the Department of Defense Electromagnetic Compatibility Analysis Center.

69. J. H. Merritt et al., "Studies on Blood-Brain Barrier Permeability after Microwave-Radiation," *Radiation and Environmental Biophysics* 15 (1978): 367–377.

70. M. H. Benedick, "Blood-Brain Barrier Workshop," Final Report on Contract N00014-79-M-0005, Office of Naval Research, 1979.

71. Don R. Justesen to James H. Merritt, January 30, 1979.

72. D. R. Justesen et al., "Introduction by the Editors" to the "Special Section on the Blood-Brain Barrier," *Radio Science* 14 (1979): 321–322; "Microwave Irradiation and the Blood-Brain Barrier," *Proceedings of the IEEE* 68 (1980): 60–67; Don Justesen to Allan Frey, March 9, 1979.

73. Benedick, "Blood-Brain Barrier Workshop," p. 5.

74. L. N. Heynick, "USAFSAM Review and Analysis of Radio Frequency Radiation Bioeffects Literature: First Report," SAM-TR-81-24 (November 1981), p. 103.

75. Justesen, "Microwave Irradiation," p. 65.

76. Don Justesen to Allan Frey, March 9, 1979.

77. Benedick, "Blood-Brain Barrier Workshop," p. 8.

78. Ibid., p. 3. Oscar's 1977 publication makes no mention of an attempt to replicate Frey's work.

79. "Blood-Brain Barrier Workshop," manuscript, March 21, 1979, p. 8/SP-441-1/A8.

80. Allan Frey, "On Microwave Effects at the Blood-Brain Barrier," *Bioelectromagnetics Society Newsletter* 18 (November 1980): 4–5; Thomas Rozzell to Allan Frey, September 15, 1980; Leonard Libber to Allan Frey, January 30, 1980.

81. Elliot Postow to Allan Frey, October 5, 1981.

82. A. S. Segal and R. L. Magin, "Microwave and the Blood-Brain Barrier: A Review," *Journal of Bioelectricity* 2 (1983): 83–88.

83. *Richmond Symposium* (1969), p. 249.

Chapter 8

1. H. Pollack, "The Microwave Syndrome," *Bulletin of the New York Academy of Medicine* 55 (December 1979): 1240–1243.

2. Barie Hartman, "The Radio Wave Story: Whys and Wherefores," *Eugene Register Guard,* April 5, 1978.

3. Gene Smith, *New York Times,* May 19, 1967.

4. Examples can be found in articles in the *New York Times,* June 11, July 8, 25, August 1, 15, 1967.

5. "Microwave Ovens," *Consumer Reports* (April 1973): 221–230.

6. Ibid., p. 222.

7. Jack Anderson, "Washington Merry-Go-Round," syndicated column, March 7, 1971. The study was an informal survey of former EC-121 pilots conducted by Russell Carpenter, who discovered twelve cases of cataracts in the sample population. Because of uncertainty about exposure conditions, however, he did not attribute the cataracts to microwave exposure and never published his initial report: "Case Reports of Effects Associated with Accidental Exposure to Microwaves."

8. Anderson, "Washington Merry-Go-Round," April 21, 1971. The study referred to is Milton Zaret, "Clinical Effects of Non-Ionizing Radiation," presented at IEEE meeting, New York, March 1971.

9. Anderson, "Washington Merry-Go-Round," June 3, 1971. The officer in question, Colonel Alvin Burner, does not attribute his transfer following the 1968 IMPI meeting, at which he did support increased bioeffects research, to his stand on RF bioeffects (private conversation).

10. Ibid., November 12, 1972.

11. Ibid., March 10, 1973.

12. Ibid., May 4, 1973.

13. Ibid., May 10, 1973.

14. Walter Stoessel to Lawrence Eagleburger, telegram, February 5, 1976, USSD-MW IV, no. 111.

15. Eagleburger to Stoessel, cable, February 6, 1976, USSD-MW IV, no. 114.

16. Stoessel to Eagleburger, telegram, February 6, 1976, USSD-MW IV, no. 113.

17. "Transcript of Press, Radio and Television News Briefing, Monday, February 9, 1976, 12:51 P.M.," USSD-MW IV, no. 126.

18. "Transcript of Press, Radio and Television News Briefing, Wednesday, February 11, 1976, 1:09 P.M.," USSD-MW IV, no. 144.

19. "The Secretary of State, Press Conference, 12 February 1976," USSD-MW IV, no. 150.

20. *Boston Globe,* February 16, 1976; see also an earlier story by Nigel Wade, "U.S. Envoy Victim of Russian Ray," *London Daily Telegraph,* February 13, 1976.

21. Thomas Love, "Radiation at U.S. Embassy Long Known, Ex-Agents Say," *Washington Star News,* February 21, 1976. Barton Reppert's stories began to appear in late April 1976.

22. "Screens Found to Block Rays at Moscow Embassy," *Washington Post,* April 26, 1976.

23. "2 Ailing U.S. Children Leave after Moscow Radiation Test," *Washington Star,* June 26, 1976.

24. "Transcript of Press, Radio and Television News Briefing, Friday, February 27, 1976, 12:48 P.M.," USSD-MW IV, no. 182.

25. "Secretary of State, Press Conference, 12 February 1976."

26. Confidential briefing paper, "Moscow Microwave Signal," circulated in early March 1976, USSD-MW IV, no. 191.

27. Operations memorandum, subject: "New Construction East of Embassy," June 29, 1973, USSD-MW II, no. 74.

28. For a description of the Moscow signals, see R. C. Mallalieu, "Microwave Intensity Distribution within the U.S. Embassy in Moscow."

29. "Subject: Ongoing Program for MUTS Measurements," cable, December 15, 1975, USSD-MW III, no. 93.

30. See, for example, S. M. Bawin, R. J. Gavalas-Medici, and W. R. Adey, "Effects of Modulated UHF Fields on Specific Brain Rhythms in Cats," *Brain Research* 58 (1973): 365–384.

31. Cover note on J. F. Schapite, "Experimental Investigation of Effectiveness of PsychoPhysiological Manipulation Using Modulated Electromagnetic Energy for Direct Information Transmission," January 1974, USSD-MW II, no. 75.

32. Eagleburger to Stoessel, telegram, January 23, 1976, USSD-MW IV, no. 108.

33. Ibid., February 4, 1976, USSD-MW IV, no. 109.

34. For Pollack's own description of his role in the Moscow embassy affair, see his testimony in Senate Hearings (1977), pp. 268–283.

35. Most of the research on the blood count problem was done in mid-1976 and early 1977 at Johns Hopkins under Dr. Abraham Lillienfeld; see USSD-MW V, nos. 286, 292, 293, 295, 296, VI, nos. 371, 374, 377, 379, 393, and VIII, no. 458.

36. Eagleburger to U.S. Delegation to Salt II in Geneva, telegram, March 17, 1976, USSD-MW IV, no. 215.

37. See the letters dated March 5, 1976, from Eagleburger to James Keogh, director, USIF; Earl Butz, secretary of agriculture; and Donald Rumsfeld, secretary of defense, USSD-MW IV, no. 207.

38. "Transcript of Press, Radio, and Television News Briefing, Wednesday, July 7, 1976," USSD-MW VI, no. 308, p. B-5, and "Soviet Dims Beam at U.S. Embassy," *New York Times,* July 8, 1976.

39. In addition to the references cited in note 35 on the blood count problem, see: for the two children evacuated, USSD-MW V, nos. 299–301, VI, nos. 306, 307; for responses to Congress, USSD-MW V, nos. 275–279, VI, nos. 316, 334, 349, 352; for dealings with AFSA, USSD-MW V, nos. 228, 230, 236, 251; and for discussions of the Moscow Viral Study, USSD-MW V, no. 297, VI, nos. 323–330. Each episode is complex in its own right and cannot be discussed in detail.

40. Paul Brodeur, "A Reporter at Large: Microwaves-I and -II," *New Yorker,* December 13, 27, 1976.

41. Brodeur, *Zapping of America.*

42. Senate Hearings (1977). The investigation into the Moscow embassy affair resulted in one brief, mildly critical report.

43. "Warning: Microwave Radiation," produced by Richard Clark, for "60 Minutes," June 19, 1977.

44. Stephen S. Rosenfeld, "Radiation Sickness: Medical and Political," *Washington Post,* January 7, 1977.

45. See reviews in: *Publisher's Weekly,* September 26, 1977, p. 129; Victor K. McElheny, "Microwaves and Men," *New York Times,* January 7, 1978; *Kirkus,* October 15, 1977, p. 1122; *Booklist,* January 1, 1978, p. 721; and Robert Claiborne, "Electronic Pollution," *Saturday Review,* January 7, 1978, pp. 35–36.

46. Claiborne, "Electronic Pollution," p. 35.

47. Senate Hearings (1977), p. 213.

48. John Osepchuk, letter to the editor, *Microwaves* (January 1978).

49. See the following letters: Allan Frey to Ivan Getting, president, IEEE, January 26, 1978; Getting to Frey, February 13, 1978; Frey to Getting, February 20, 1978; Frey to Getting, February 27, 1978; Getting to Frey, March 31, 1978; Richard Damon, director, Division IV, IEEE, to Frey, April 27, 1978; Frey to Damon, May 22, 1978.

50. D. R. Justesen and C. Susskind in *IEEE Spectrum* 15 (May 1978): 60–63.

51. Anon. letter to the editor, *Microwave Systems News* (May 1978): 10.

52. For other critical reviews, see William Herman, "Potential Hazards," *Science* 200 (May 1978): 643–645 and Victor McElheny, "Microwaves and Men," *New York Times,* January 7, 1978.

53. "Microwave Effects Worry Some Scientists," *Hartford Courant,* February 27, 1977; Stephen J. Lynton, "Microwave Radiation Health Hazard Worries Scientists," *Washington Post,* July 10, 1977; Ron Chernow, "Debating the Microwave Danger," *Philadelphia Inquirer,* September 18, 1977; "The Zapping of America by Paul Brodeur," *Eugene Register-Guard,* April 23, 1978. These citations represent only a small fraction of the news stories that have appeared since early 1977.

54. See *People's Weekly,* January 30, 1978, pp. 25–27; *New Times,* March 6, 1978, pp. 29–37, 60–64; *Newsweek,* July 17, 1978; *Time,* August 28, 1978; *Saturday Review,* September 15, 1979; *Science 80* (1980); *Science for the People* (March–April 1980); *Let's Live* (February 1980); *Environment* 12, nos. 4 and 5 (May and June 1970); *New York,* June 9, 1980; *Reader's Digest* (January 1980).

55. For background on the Vernon Valley case, see the series of articles by Chapin Wright, *Trenton Sunday Times,* December 26, 1982.

56. See Hartman, "Radio Wave Story."

57. Brodeur, *Zapping of America,* p. 183; *People's Weekly,* January 30, 1978, p. 27.

58. Brodeur, *Zapping of America,* p. 324.

59. Ibid., p. 117.

60. Richard Cesaro to Director, Defense Research and Engineering, top secret memorandum, "Project Bizarre," September 27, 1967.

61. Brodeur, *Zapping of America,* p. 115.

62. Ibid., p. 105.

63. Ibid., p. 129.

64. Ibid., p. 133; Pandora, minutes, April 21, 1969, p. 4.

65. Brodeur, *Zapping of America,* p. 129.

66. Ibid., p. 292.

67. Senate Hearings (1973), p. 19.

68. Brodeur, *Zapping of America,* p. 73.

69. Robert L. Elder to James Frazer, June 20, 1976. This letter was circulated to ANSI members on October 26, 1976, along with other "Recommendations concerning the future status of the C95.1-1966 Standard."

70. Sam Koslov, William Mills, David Donaldson, Sol Michaelson, and John Osepchuk to Harry Sonnemann, Special Assistant to the Assistant Secretary of the Navy (R&D), Subject: "Review of Contract N00014-69-C-0358, 'Radio-Frequency Hazards,' Principal Investigator: Dr. Milton M. Zaret," August 16, 1971.

71. Milton Zaret, "Absence of Heart Rate Effects in Rabbits during Low Level Microwave Irradiation," *IEEE Transactions on Microwave Theory and Technique* (February 1971): 168–173.

72. See "Combatting the Statistics of Heart Disease," *Washington Post,* April 17, 1977. Zaret's suggestions about the microwave link were given in "Electronic Smog as a Potentiating Factor in Cardiovascular Disease: A Hypothesis of Microwaves as an Etiology for Sudden Death from Heart Attack in North Karelia," *Medical Research Engineering* 12 (1976).

73. *Reader's Digest* (January 1980): 68.

74. *Saturday Review,* September 15, 1979, p. 16.

75. *Saturday Review,* January 7, 1978, pp. 35–36.

Chapter 9

1. For a survey of legal actions, see *Microwave News* (April 1981, December 1982).

2. "Temporary Consensus Standards on Nonionizing Radiation," issued under the Occupational Safety and Health Act of 1970, published in *Federal Register,* May 19, 1971, pp. 10,522–10,523.

3. OSHA, *Occupational Safety and Health Decisions* (1975–1976), no. 20,379.

4. Robert F. Cleveland, Jr., "Opinions on Costs, Benefits, and Risks: Results of an Inquiry by the Federal Communications Commission on Potential Hazards of Radio-frequency Radiation," in Steneck, *Risk/Benefit,* pp. 69–96.

5. Morris L. Greb, attorney for Rockaway Township Zoning Board of Adjustment, to ANSI, August 5, 1980.

6. See the coverage of the application process in *Smithtown News,* November 12, 19, 26, 1981, January 7, 14, 1982, February 11, 1982; *Smithtown Messenger,* November 5, 12, 19, 1981, January 7, 14, 1982; and *New York Times,* December 6, 1981. The variance was awarded on February 11, 1982, case 7145, Office of the Board of Zoning Appeals, Town of Smithtown.

7. Tim Welch, "RCA Earth Station: Good Neighbor or Trouble?" *Bainbridge Review,* July 23, 1981.

8. Molly Ugles, president, BHNA, to Rick Kimball, SEPA Coordinator, July 25, 1981; William D. Rives, of Davis, Wright, Todd, Riese, and Jones, to Rick Kimball, August 19, 1981.

9. A. W. Guy to Rick Kimball, August 19, 1981.

10. The county's decision was communicated in a notice dated August 21, 1981, Kitsap County Department of Community Development. The notice was signed by Rick Kimball for the department's director, Ron Perkerewicz.

11. *Bainbridge Review,* August 26, 1981.

12. David A. Bricklin of the law offices of Roger M. Leed, to Ron Perkerewicz and Rick Kimball, October 14, 1981.

13. Ibid.

14. Battelle Human Affairs Research Center, "Draft Environmental Impact Statement, RCA Earth Station, Bainbridge Island, Washington" (March 1982).

15. Bucklin Hill Neighborhood Association et al., "Response to RCA Earth Station Draft E.I.S.," April 1982.

16. Tim Welch, "Microwave Opponents Blast EIS; Criticize Its Authors," *Bainbridge Review,* April 28, 1982.

17. Roger Tilton, "UW Prof Is Consultant: Bainbridge Microwave Station Ignites Fear," *Daily,* February 18, 1982; Tim Welch, "UW Researcher Says Microwave Controversy Fueled by Misinformation," *Bainbridge Review,* March 31, 1982.

18. John R. Spencer, Region X administrator, EPA, to Rick Kimball, May 12, 1982.

19. Richard A. Tell to Gary L. O'Neal, director, Environmental Services Division, EPA, Washington State, April 20, 1982, and Joe A. Elder to Dick Thiel, chief, Environmental Evaluation Branch, Region X, April 15, 1982.

20. "Vashon Island Now Fighting Microwaves," *Bainbridge Review,* November 11, 1981; "No Decision Yet: RCA Pondering NK Microwave Sites," *Kitsap County Herald,* January 13, 1982.

21. William Rives, of Davis, Wright, Todd, Riese, and Jones, to Rick Kimball, June 28, 1982; Ron Lamb, "RCA Switches Sites for Microwave Unit," *Bremerton Sun,* June 30, 1982; and Tim Welch, "RCA Asks for Deferral on Bainbridge Microwave," *Bainbridge Review,* June 30, 1982.

22. "Planning Department Report: Site Location Review," Kitsap County Planning Department, July 15, 1982.

23. Leonard Costello, "Findings of Fact, Conclusions and Recommendations to the Kitsap County Board of Commissioners," no. 820720153, September 2, 1982.

24. Ibid., p. 17.

25. John Horsley, commissioner, to Kitsap County Board of Commissioners, memorandum, "RCA American Communications Inc., VPUP," October 18, 1982, p. 1.

26. Ibid., p. 3.

27. Ibid.

28. Several papers in Steneck, *Risk/Benefit,* address risk assessment from a scholarly point of view. See especially R. A. Albanese and M. L. Winfree, "Approaches to Risk Assessment: A Suggested Use for Cost/Benefit Analysis in the Microwave Debate."

29. *Joseph F. Kerch* v. *Air America, Inc.,* U.S. Department of Labor, Office of Administrative Law Judges, case 15-19092/80-LHCA-1388 and 15-16930/80-LHCA, filed January 5, 1978.

30. *Microwave News* (December 1982): 3.

31. New York State Workers' Compensation Claims nos. 0714-2308, 0752-3602.

32. For a summary of the witnesses, see Louis Slessin, "Microwave Death Award Upheld in Compensation Court, NY Tel Appeals Again," *Microwave News* (April 1981): 1, 4–6.

33. Ibid., p. 6.

34. Ibid. (June 1982): 2.

35. Ibid. (July–August 1982): 2–3.

36. For general descriptions of the two cases, see articles in *Providence Journal-Bulletin,* March 18, 22, June 14, 1977, and by Nancy Pappus in *Hartford Courant,* March 19, 1977.

37. Case TEC-0506, Twenty-eighth U.S. District Court, Hartford, Connecticut, filed March 17, 1977.

38. Stephen Morin and Wayne Worcester, "Quonset Worker Suit Ties Microwave to Cancer," *Providence Journal-Bulletin,* March 18, 1977.

39. Hamilton F. Allen, "Widow Says Radiation Caused Husband to Die," *Providence Journal-Bulletin,* June 14, 1977. Her case was barred under the statute of limitations of the state in which the suit was filed, Rhode Island; see *Microwave News* (January–February 1983): 5.

40. See Jonathan Winer, "They Call Themselves the 'Victims Network,'" *Boston Globe,* July 25, 1976; Joe Towne, memorandum, "'RADVICNET', Radar Victims Network," n.d.; Towne to Claire Odgers, *London Telegraph Sunday Magazine,* April 5, 1978.

41. For a listing of documents filed, see *Engell* v. *International Telephone & Telegraph Corp. et al.,* Civil H-77-130, U.S. District Court, Hartford, Connecticut.

42. *Microwave News* (January–February 1983): 4–5.

43. Ibid. (December 1982): 1–4.

44. "Deposition of Milton M. Zaret" in *Mulhauser and Ryan* v. *Litton Industries,* Superior Court of New Jersey, Law Division: Essex County, Docket No. L-29107-72, October 2, 1975.

45. Ibid., esp. pp. 32–34, 268–283.

46. Ibid., pp. 107–108.

47. "Civil Action Stipulation of Dismissal with Prejudice," *Mulhauser-Ryan* v. *Litton Industries,* Superior Court of New Jersey, Law Division, Essex County, docket L-29197-72, January 9, 1976.

48. *HBO Hearings,* VIII:64–65.

Chapter 10

1. House Hearings (1981).

2. See *Concepts and Approaches for Minimizing Excessive Exposure to Electromagnetic Radiation from RF Sealers,* prepared by BRH, FDA 82-8192 (1982).

3. American National Standard: *Safety Levels with Respect to Human Exposure to Radio Frequency Electromagnetic Fields, 300 kHZ to 100 GHz* (New York: IEEE, 1982) (hereafter cited as C95.1-1982).

4. A more detailed discussion of the argument that follows can be found in N. H. Steneck, "Subjectivity in Standards: The Case of ANSI C95.1-1982," *Microwaves and RF* (May 1983): 137ff.

5. C95.1-1982, par. 5.

6. Ibid., par. 6.5.

7. Ibid., par. 6.3.

8. Ibid., par. 6.2.

9. A. W. Guy, M. D. Webb, and C. C. Sorensen, "Determination of Power Absorption in Man Exposed to High-Frequency Electromagnetic Fields by Thermographic Measurements on Scale Models," *IEEE Transactions on Biomedical Engineering* 23 (1976): 361–371.

10. C95.1-1982, par. 6.2.

11. Ibid., par. 6.6.

12. Steneck, "Subjectivity in Standards," and "The Relationship of History to Policy," *Science, Technology, and Human Values* 7 (1982): 105–112.

Bibliographic Note

This brief literature survey is intended to provide general suggestions for additional reading. Those with particular interests can find other sources listed in the bibliographic and general survey works mentioned below and in the works cited in the notes in this book. Manuscript and technical materials used in preparing this book have not been listed in a separate bibliography; full references have been given in the notes and can be consulted as needed.

The best source of current information on the microwave debate is *Microwave News: A Monthly Report on Non-Ionizing Radiation,* ed. Louis Slessin, PO Box 1799, Grand Central Station, NY 10163. At $200 for a year's subscription, this publication is most suited to the needs of groups that have an interest in developments on all fronts. Professionals in bioeffects research can get a much more limited perspective on current developments through the Bioelectromagnetics Society's *Newsletter,* 1 Bank Street, Gaithersburg, MD 20760.

Information on current scientific publications on bioeffects research and related developments is given in *Biological Effects of Nonionizing Electromagnetic Radiation: A Digest of Current Literature,* a quarterly publication produced for the Office of Naval Research by Information Ventures, 1500 Locust Street, Philadelphia, PA 19102. The best cumulative index of scientific publications in this field is Zorach R. Glaser and Julia L. Moore, *Cumulative Index to the Bibliography of Reported Biological Effects and Clinical Symptoms Attributed to Microwave and Radio-Frequency Radiation Exposure* (Riverside, CA: Julie Moore & Associates, 1982). This index, which is based on a series of earlier bibliographies compiled by Glaser, contains many listings of nonscientific literature that are of general interest.

Most general survey works pertaining to the microwave debate have focused on scientific publications and must be read with caution since they usually have been prepared for specific purposes. The most comprehensive surveys are: National Institute of Occupational Safety and Health, *RF/Microwave Criteria Document: External Review Draft,* 2 vols. (n.d., circulated 1979–1980); and Environmental Protection Agency, *Biological Effects of Radiofrequency Radiation: External Review Draft,* no. EPA–600/8–83–026A (June 1983). Om P. Gandhi, "Biological Effects and Medical Applications of RF Electromagnetic Fields," *IEEE Transactions on Microwave Theory and Techniques* 30 (1982): 1831–1847, provides a much briefer survey of the scientific literature, which follows in a long line of summaries produced by the community doing the research. S. Baranski and P. Czerski, *Biological Effects of Microwaves* (Stroudsburg, PA: Dowden, Hutchinson & Ross, 1976), has extensive coverage of the Soviet and East European literature, which is still useful even though it is now out-of-date. Many articles in *Symposium on Health Aspects of Nonionizing Radiation,* published in *Bulletin of the New York Academy of Medicine* 55 (December 1979), have useful summaries, although the quality is mixed.

Balanced surveys of nonscientific issues are limited in number. Karen Massey, "The Challenge of Nonionizing Radiation: A Proposal for Legislation," *Duke Law Journal* 105 (1979): 105–188, gives an excellent survey of prior legislative activities. Leonard Davis, "A Study of Federal Microwave Standards," a report prepared for the U.S. Department of Energy, Office of Energy Research, DOE/ER/10041–02 (August 1980), covers similar material more briefly and with greater attention to standards. N. H. Steneck, H. J. Cook, A. J. Vander, G. L. Kane, "The Origins of U.S. Safety Standards for Microwave Radiation," *Science* 208 (June 1980): 1230–1237, sketches out the history of RF standards through the adoption of ANSI C95.1–1966.

Among popular writings Paul Brodeur, *The Zapping of America: Microwaves, Their Deadly Risk, and the Cover-up* (New York: W. W. Norton, 1977), still provides the most readable general account of the development of the microwave debate. This work is not always reliable, and readers should exercise caution in using both Brodeur's information and his interpretations. Most subsequent popular accounts have been heavily influenced by Brodeur's cover-up thesis.

Finally, for readers who want to get firsthand a feeling for the different opinions that have led to the microwave debate, a ponderous but sometimes entertaining place to begin is with the reoccurring congressional hearings: U.S. House, Committee on Interstate and Foreign Commerce, Subcommittee on Public Health and Welfare, *Electronic Products Radiation Control, Hearings on H.R. 10790,* 90th Congress, 1st sess., 1967; U.S. Senate, Committee on Commerce, *Radiation Control for Health and Safety Act of 1967, Hearings on S. 2067, S. 3211, and H.R. 10790,* 90th Congress, 1st, 2nd sess., 1967–1968; U.S. Senate, Committee on Labor and Public Welfare, Subcommittee on Health, *Radiation Health and Safety Act, 1974, Hearings on S. 667,* 93rd Congress, 2nd sess., 1974; U.S. Senate, Committee on Commerce, Science, and Transportation, *Radiation Health and Safety, Hear-*

ings on Oversight of Radiation Health and Safety, 95th Congress, 1st sess., 1977; U.S. House, Committee on Science and Technology, Subcommittee on Natural Resources and Environment, *Research on Health Effects of Nonionizing Radiation, Hearing,* 96th Congress, 1st sess., 1979; and U.S. House, Committee on Science and Technology, Subcommittee on Investigations and Oversight, *Potential Health Effects of Video Display Terminals and Radio Frequency Heaters and Sealers, Hearings,* 97th Congress, 1st sess., 1981.

Index